QCE Units 3 & 4
PHYSICS

SCOTT ADAMSON

T0362514

+ topic summary notes
+ exam practice questions
+ detailed, annotated solutions
+ study and exam preparation advice

STUDY NOTES

A+ Physics QCE Units 3 & 4 Study Notes Workbook
1st Edition
Scott Adamson
ISBN 9780170459174

Publisher: Sam Bonwick
Project editor: Alan Stewart
Cover design: Nikita Bansal
Text design: Alba Design
Project designer: Nikita Bansal
Permissions researcher: Wendy Duncan
Production controller: Jaimi Kuster
Typeset by: SPi Global

Any URLs contained in this publication were checked for currency during the production process. Note, however, that the publisher cannot vouch for the ongoing currency of URLs.

Acknowledgements

Definitions (c) State of Queensland (QCAA)
Physics General Senior Syllabus 2019
CC BY 4.0 https://creativecommons.org/licenses/by/4.0/

© 2021 Cengage Learning Australia Pty Limited

Copyright Notice

This Work is copyright. No part of this Work may be reproduced, stored in a retrieval system, or transmitted in any form or by any means without prior written permission of the Publisher. Except as permitted under the *Copyright Act 1968,* for example any fair dealing for the purposes of private study, research, criticism or review, subject to certain limitations. These limitations include: Restricting the copying to a maximum of one chapter or 10% of this book, whichever is greater; providing an appropriate notice and warning with the copies of the Work disseminated; taking all reasonable steps to limit access to these copies to people authorised to receive these copies; ensuring you hold the appropriate Licences issued by the Copyright Agency Limited ("CAL"), supply a remuneration notice to CAL and pay any required fees. For details of CAL licences and remuneration notices please contact CAL at Level 11, 66 Goulburn Street, Sydney NSW 2000,
Tel: (02) 9394 7600, Fax: (02) 9394 7601
Email: info@copyright.com.au
Website: www.copyright.com.au

For product information and technology assistance,
in Australia call **1300 790 853**;
in New Zealand call **0800 449 725**

For permission to use material from this text or product, please email
aust.permissions@cengage.com

ISBN 978 0 17 045917 4

Cengage Learning Australia
Level 7, 80 Dorcas Street
South Melbourne, Victoria Australia 3205

Cengage Learning New Zealand
Unit 4B Rosedale Office Park
331 Rosedale Road, Albany, North Shore 0632, NZ

For learning solutions, visit **cengage.com.au**

Printed in China by 1010 Printing International Limited.
1 2 3 4 5 6 7 25 24 23 22 21

CONTENTS

UNIT 3

GRAVITY AND ELECTROMAGNETISM

REVOLUTIONS IN MODERN PHYSICS

UNIT 4

9780170459174

HOW TO USE THIS BOOK

A+ Physics QCE Units 3 & 4 Study Notes is designed to be used year-round to prepare you for your QCE Physics exam. *A+ Physics QCE Units 3 & 4 Study Notes* includes topic summaries of all the key knowledge in the A+ Physics QCE syllabus that you will be assessed on during your exam. Chapters 1 & 2 and 4–6 address each of the topics in the syllabus. Chapter 3 specifically addresses the Data test, which is a separate assessment task worth 10% of your final mark. This section gives you a brief overview of each chapter and the features included in this resource.

Topic summaries

The topic summaries at the beginning of each chapter give you a high-level summary of the essential subject matter for your exam.

Concept maps

The concept maps at the beginning of each topic provide a visual summary of the hierarchy and relationships between the subject matter in each topic.

Key knowledge summaries

Key knowledge summaries in each chapter address all key knowledge of the syllabus. Summaries are broken down into sequentially numbered chunks for ease of navigation. Step by step worked examples and hints unpack the content and prevent mistakes.

Glossary

All your key terms for each topic are bolded throughout the key knowledge summaries and included in a complete glossary near the end of each topic. Digital flashcards are accessible via QR code and provide a handy revision tool.

Revision summary

The revision summaries are a place for you to make notes against each of the syllabus dot points and ensure you have thoroughly reviewed the content.

Exam practice

Each topic ends with multiple-choice and short response questions for you to test your recall of the key concepts and practise answering the types of questions you will face in your exam. Complete solutions to practice questions are available at the back of the book to provide immediate feedback and help self-correct errors. They have been written to reflect a high-scoring response. This section includes explanations of why the multiple-choice answers are correct, and explanations for short response items that demonstrate what a high-scoring response looks like, with mark breakdowns, and signpost potential mistakes.

PREPARING FOR THE EXAM

Exam preparation is a year-long process. It is important to keep on top of the theory and consolidate little and often, rather than leaving work to the last minute. You should aim to have the theory learned and your notes complete so that by the time you reach study leave, the revision you do is structured, efficient and meaningful.

Study tips

To stay motivated to study, try to make the experience as comfortable as it can be. Have a dedicated study space that is well lit and quiet. Create and stick to a study timetable, take regular breaks, reward yourself with social outings or treats and use your strengths to your advantage. For example, if you are a visual learner, turn your Physics notes into cartoons, diagrams, or flow charts. If you are better with words or lists, create flash cards or film yourself explaining tricky concepts and watch it back.

Revision techniques

Here are some useful revision methods to help information **STIC**.

Spaced repetition	This technique uses the Leitner method, which helps to move information from your short-term memory into your long-term memory by spacing out the time between when you are asked to revise or recall information from flash cards you have created. As the time between retrieving information is slowly extended, the brain processes and stores the information for longer periods.
Testing	Testing is necessary for learning and is a proven method for exam success. The 'hypercorrection effect' shows when you are confident of an answer which is actually incorrect, you are more likely to remember the correct answer, thereby improving your future performance. Further, if you test yourself before you learn all the content, your brain becomes primed to retain the correct answer when you get it.
Interleaving	A revision technique that sounds counterintuitive but is very effective for retaining information. Most students tend to revise a single topic in a session, and then move onto another topic next session. With interleaving, you may choose three topics (1, 2, 3) and spend 20-30 minutes on each topic. You may choose to study 1, 2, 3 or 2, 1, 3 or 3, 1, 2 'interleaving' the topics, and repeating the study pattern over a long period of time. This strategy is most helpful if the topics are from the same subject and are closely related.
Chunking	An important strategy is breaking down large topics into smaller, more manageable 'chunks' or categories. Essentially, you can think of this as a branching diagram or mind map where the key theory or idea has many branches coming off it which get smaller and smaller. By breaking down the topics into these chunks, you will be able to revise the topic systematically.

These strategies take cognitive effort, but that is what makes them much more effective than re-reading notes or trying to cram information into your short-term memory the night before the exam!

Time management

It is important to manage your time carefully throughout the year. Make sure you are getting enough sleep, that you are getting the right nutrition, and that you are exercising and socialising to maintain a healthy balance so that you don't burn out.

To help you stay on target, plan out a study timetable. One way to do this is to:

1. Assess your current study time and social time. How much are you dedicating to each?
2. List all your commitments and deadlines, including sport, work, assignments, etc.
3. Prioritise the list and re-assess your time to ensure you can meet all your commitments.
4. Decide on a format, whether it be weekly or monthly, and schedule in a study routine.
5. Keep your timetable somewhere you can see it.
6. Be consistent.

The exam

The end-of-year examination accounts for 50% of your total mark. It assesses your achievement in the following objectives of units 3 & 4:

1. describe and explain gravity and motion, electromagnetism, special relativity, quantum theory and the Standard Model
2. apply understanding of gravity and motion, electromagnetism, special relativity, quantum theory and the Standard Model
3. analyse evidence about gravity and motion, electromagnetism, special relativity, quantum theory and the Standard Model to identify trends, patterns, relationships, limitations or uncertainty
4. interpret evidence about gravity and motion, electromagnetism, special relativity, quantum theory and the Standard Model to draw conclusions based on analysis.

The examination includes two papers. Each paper features:
- multiple-choice questions
- short response items requiring single-word, sentence or paragraph responses
- calculating using algorithms
- interpreting graphs, tables or diagrams
- responding to unseen data and/or stimulus.

You will have 90 minutes plus 10 minutes perusal time for each paper.

Modified from Physics General Senior Syllabus 2019,
© State of Queensland (QCAA) 2019, licensed under CC BY 4.0

The day of the exam

The night before your exam, try to get a good rest and avoid cramming, as this will only increase stress levels. On the day of the exam, arrive at the venue of your exam early and bring everything you will need with you. If you have to rush to the exam, you will increase your stress levels, thereby lowering your ability to do well. Further, if you are late, you will have less time to complete the exam, which means that you may not be able to answer all the questions or may rush to finish and make careless mistakes. If you are more than 30 minutes late, you may not be allowed to enter the exam. Don't worry too much about 'exam jitters'. A certain amount of stress is required to help you concentrate and achieve an optimum level of performance. If, however, you're still feeling very nervous, breathe deeply and slowly. Breathe in for a count of six seconds, and out for six seconds until you begin to feel calm.

Perusal time

Use your time wisely! *Do not* use the perusal time to try and figure out the answers to any of the questions until you've read the whole paper! The exam will not ask you a question testing the same knowledge twice, so look for hints in the stem of the question and avoid repeating yourself. Plan your approach so that when you begin writing you know which section, and ideally which question you are going to start with.

Strategies for effective responses

Pay particular attention to the cognitive verb used in the question. For example, a question with the cognitive verb 'explain' requires a different response to a question with the cognitive verb 'describe'. Familiarise yourself with the definitions of the commonly used cognitive verbs below (listed in order of complexity). Understanding the definitions of these cognitive verbs will help ensure you are not just providing general information or restating the question without answering it.

describe	give an account (written or spoken) of a situation, event, pattern or process, or of the characteristics or features of something
explain	make an idea or situation plain or clear by describing it in more detail or revealing relevant facts; give an account; provide additional information
apply	use knowledge and understanding in response to a given situation or circumstance; carry out or use a procedure in a given or particular situation
analyse	dissect to ascertain and examine constituent parts and/or their relationships; break down or examine in order to identify the essential elements, features, components or structure; determine the logic and reasonableness of information; examine or consider something in order to explain and interpret it, for the purpose of finding meaning or relationships and identifying patterns, similarities and differences
interpret	use knowledge and understanding to recognise trends and draw conclusions from given information; make clear or explicit; elucidate or understand in a particular way; bring out the meaning of, e.g. a dramatic or musical work, by performance or execution; bring out the meaning of an artwork by artistic representation or performance; give one's own interpretation of; identify or draw meaning from, or give meaning to, information presented in various forms, such as words, symbols, pictures or graphs

Physics General Senior Syllabus 2019, © State of Queensland (QCAA) 2019, licensed under CC BY 4.0

Write down all the key steps to show your working when calculating answers. Each mark awarded for a question relates to a step in the calculation. Your *A+ Physics Study Notes* book features worked examples and step-by-step worked solutions to help familiarise you with these types of questions.

Multiple-choice questions

Read the question carefully and underline any important information to help you break the question down and avoid misreading it. Read all the possible solutions and eliminate any clearly incorrect answers. Fill in the multiple-choice answer sheet carefully and clearly. Check your answer and move on. Do not leave any answers blank.

Short response questions

It is important that you plan your response before writing. To do this, **BUG** the question:
- **Box** the cognitive verb (describe, explain, apply, etc).
- **Underline** any key terms or important information and take note of the mark allocation.
- **Go** back and read the question again.

Many questions require you to apply your knowledge to unfamiliar situations so it is okay if you have never heard of the context before, but you should know which part of the syllabus you are being tested on and what the question is asking you to do. If there is a stimulus included, use information from that as part of the response to show how you are linking the (unfamiliar) context to your knowledge.

Plan your response in a logical sequence. If the question says, 'describe and explain' then structure your answer in that order. You can use dot points to do this, but ensure you write in full sentences. Rote-learned answers are unlikely to receive full marks, so you must relate the concepts of the syllabus back to the question and ensure that you answer the question that is being asked. Planning your response to include the relevant information and the key terminology will help you from writing too much, contradicting yourself, or 'waffling' on and wasting time. If you have time at the end of the paper, go back and re-read your answers.

Good luck. You've got this!

ABOUT THE AUTHOR

Scott Adamson

Scott Adamson is an experienced Science and Mathematics author, reviewer, and Curriculum Leader at an independent girls' school in Brisbane. He is currently holds several QCAA Assessor roles in Physics, including Lead Confirmer and Chief External Marker, and was a member of the Learning Area Reference Group for the introduction of the new QCE in Queensland. Scott was lead author on the *Nelson QScience Physics* series.

A+ DIGITAL FLASHCARDS

Revise key terms and concepts online with the A+ Flashcards. Each topic glossary in this book has a corresponding deck of digital flashcards you can use to test your understanding and recall. Just scan the QR code or type the URL into your browser to access them.

https://get.ga/aplus-qce-phys-u34

Note: You will need to create a free NelsonNet account.

9780170459174

UNIT 3
GRAVITY AND ELECTROMAGNETISM

Chapter 1
Topic 1: Gravity and motion

Topic summary

The topic 'Gravity and motion' introduces scalar and vector values, resolving vectors into perpendicular components, and performing vector analysis. Projectile motion studies objects moving in two dimensions near the surface of Earth, where the gravitational field acts vertically and air resistance is considered negligible.

Inclined planes are simple machines often used as wedges and levers. They invoke components of gravitational force and applied forces such as friction and the normal force. Circular motion refers to objects travelling at constant speeds but changing direction as a result of a net force (and net acceleration) acting toward the centre of a circular path.

Gravitational fields are the basis of the universal law of gravitation as well as explaining aspects of the nature of orbits, including Kepler's three laws of planetary motion.

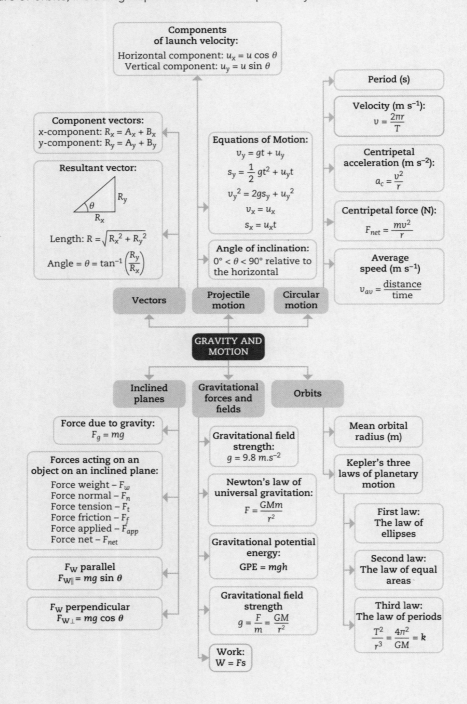

1.1 Vectors

1.1.1 Vector and directional analysis

When motion is described in kinematics, the following variables can be stated as vectors. They have both magnitude and direction.

- displacement
- velocity
- acceleration
- force
- momentum

Magnitude is stated using SI units (standard or derived):

Quantity	Unit and symbol
displacement	metres (m)
velocity	metres per second (m s^{-1})
acceleration	metres per second squared (m s^{-2})
force	newtons (N) or (kg m s^{-2})
momentum	kilogram metres per second (kg m s^{-1})
mass	kilograms (kg)
time	seconds (s)

Direction is stated using one of the agreed axis systems:

- compass points (quadrant or true bearings)
- Cartesian axes
- positive x-axis aligned with one of the vectors
- positive x-axis aligned with an inclined plane.

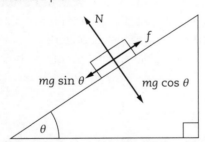

FIGURE 1.1 Forces acting on an object on an inclined plane are aligned with the plane of the sloping surface. $mg \sin \theta$ and f are parallel to the plane; N and $mg \cos \theta$ are perpendicular to the plane.

1.1.2 Vector geometric addition in two dimensions: scale drawing

Vectors can be added and subtracted geometrically. Solutions can be found by careful scale drawing using Cartesian graph paper, protractor, ruler and a fine-point pencil. Although not strictly required, it is very useful to draw a quick scale diagram to ensure that your final solution has both the correct magnitude and direction.

Tip-to-tail method

Any number of vectors may be added using the tip-to-tail method. In this example, two vectors are added by forming two sides of a triangle.

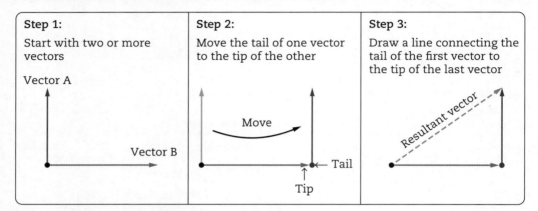

FIGURE 1.2 Adding vectors using the tip-to-tail method. The tail of vector $\vec{B}$ is added from the tip of vector $\vec{A}$. The magnitude and direction of the resultant vector AB may then be determined.

Step 1: Start with two or more vectors	Step 2: Move the tail of one vector to the tip of the other	Step 3: Draw a line connecting the tail of the first vector to the tip of the last vector
Vector A Vector B	Move Tail Tip	Resultant vector

FIGURE 1.3 Applying the tip-to-tail method in three easy steps.

1.1.3 Vector subtraction

Hint
Subtracting a vector is equivalent to adding the negative of the vector.

As in ordinary subtraction, vector subtraction is the addition of the negative. Thus:

$$\vec{C} = \vec{A} - \vec{B}$$
$$\Rightarrow \vec{C} = \vec{A} + (-\vec{B})$$

1.1.4 Scalar multiplication of vectors

A vector may be multiplied by a scalar quantity, known as a **scalar multiplier**.

Multiplication by a positive scalar changes the magnitude but not the direction of the vector.

Multiplication by a negative scalar changes the magnitude *and* reverses the direction of the vector.

Division of a vector by a scalar is the same as multiplication of the vector by the inverse of the divisor.

1.1.5 Components of vectors

Any vector can be constructed as the sum of two other vectors. Each of these two vectors is a component of the original vector. When a vector is resolved into two components, each component is called a **resolute**.

It is very useful to resolve vectors into **rectangular components** that are perpendicular to each other. This enables the use of the geometry and trigonometry of right-angle triangles. Figure 1.4 shows the rectangular resolutes of the displacement vector 55 m, N37°E. The resolutes are taken in the north and east directions respectively but could similarly be stated in the *x* and *y* directions.

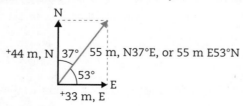

FIGURE 1.4 Resolutes of the vector 55 m, N37°E. The north component = +44 m; the east component = +33 m.

On a Cartesian grid, resolutes are taken with respect to the *x*- and *y*-axes. The angle is taken with respect to the positive direction of the *x*-axis. Resolutes may also be described using compass points (quadrant or true bearings), may be aligned with one of the vectors involved or may be aligned with the plane of a sloping surface (parallel with or perpendicular to the surface of an inclined plane).

1.1.6 Determining the resultant vector using perpendicular components

Resolutes can be added algebraically in the *x* and *y* directions because they are perpendicular. This enables the magnitude of the resultant to be calculated by Pythagoras' theorem, and the angle to be found by applying the tangent ratio.

Consider vector $\vec{A}$, which is oriented at an angle θ to the positive *x*-axis. The perpendicular components are A_x and A_y.

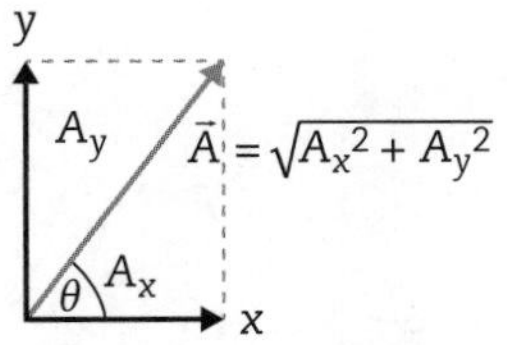

FIGURE 1.5 Vector $\vec{A}$ has perpendicular components A_x and A_y. The angle is taken relative to the positive direction of the *x*-axis.

Rectangular components of vector $\vec{A}$:

$$x\text{-component: } A_x = A \cos \theta$$

$$y\text{-component: } A_y = A \sin \theta$$

Determination of angle:

$$\tan \theta = \frac{y\text{-component}}{x\text{-component}}$$

$$\Rightarrow \theta = \tan^{-1}\left(\frac{y\text{-component}}{x\text{-component}}\right)$$

Determination of magnitude (using Pythagoras' theorem):

$$A = \sqrt{A_x^{\,2} + A_y^{\,2}}$$

Use the following general procedure for adding or subtracting two or more vectors using perpendicular components.

1 Sketch a diagram that clearly shows the vector addition.
2 Choose perpendicular *x*- and *y*-axes. Sometimes the choice is obvious, as in the case of using the compass points. Sometimes it is better to select the *x*-axis to be along one of the vectors so that vector has a single component.
3 Resolve each vector into perpendicular resolutes (*x*- and *y*-components, compass points, etc.). Be sure to identify positive and negative values for the resolutes.
4 Calculate the magnitude of the total *x*-component.
5 Calculate the magnitude of the total *y*-component.
6 Find the magnitude of the resultant using Pythagoras' theorem.
7 Find the angle using the tangent ratio.
8 Check that the angle is given in the terms required by the question.

Worked example

A barge is pulled through the water by two boats. Boat A applies a tension force of 4.0×10^4 N E30°N. Boat B pulls with a tension force of 6.0×10^4 N E50°S. Determine the resultant force applied by the boats.

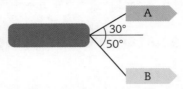

FIGURE 1.6 Free-body diagram showing forces applied by two boats pulling a barge.

Answer

x-components:

The resultant will be east as both boats pull towards the east.

⇒ take east as positive *x*-direction.

$$R_x = A_x + B_x$$

$$= 4.0 \times 10^4 \text{N} \times \cos 30° + 6.0 \times 10^4 \text{N} \times \cos 50°$$

$$= 3.464 \times 10^4 \text{N} + 3.857 \times 10^4 \text{N}$$

$$= 7.320 \times 10^4 \text{N East}$$

y-components:

The resultant is likely to be more towards the south because Boat B pulls with a force of greater magnitude.

⇒ take south as positive *y*-direction:

$$R_y = A_y + B_y$$

$$= -4.0 \times 10^4 \text{N} \times \sin 30° + 6.0 \times 10^4 \text{N} \times \sin 50°$$

$$= -2.000 \times 10^4 \text{N} + 4.596 \times 10^4 \text{N}$$

$$= 2.596 \times 10^4 \text{N South}$$

Magnitude of resultant:

$$R = \sqrt{R_x^2 + R_y^2}$$

$$= \sqrt{(7.320 \times 10^4) + (2.596 \times 10^4)^2}$$

$$= 7.767 \times 10^4 \text{N}$$

$$= 7.8 \times 10^4 \text{N to 2 significant figures}$$

Angle:

Let θ be the angle from the horizontal *x*-axis:

$$\theta = \tan^{-1}\left(\frac{R_y}{R_x}\right)$$

$$= \tan^{-1}\left(\frac{2.596 \times 10^4}{7.320 \times 10^4}\right)$$

$$= 19.5°$$

$$= \text{East 19.5° South}$$

Quadrant or true bearing:

Let θ be the angle with respect to east direction (090°).

$\theta = 19.5°$ below the east direction.

True bearing:

$$\text{True bearing} = 090° + 19.5°$$

$$= 109.5°$$

1.2 Projectile motion

1.2.1 Horizontal and vertical components

When a projectile is launched, it is measured in terms of the launch speed, u, and the angle relative to the horizontal, θ.

The vertical component of the launch velocity is independent of the horizontal. It is measured using $u_y = u \sin \theta$.

The horizontal component is measured using $u_x = u \cos \theta$.

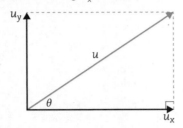

FIGURE 1.7 A projectile is launched at speed u and angle θ to the horizontal. The horizontal and vertical components of the launch velocity, u_x and u_y respectively, are shown.

Hint

Key formula

Components of the launch velocity

Horizontal component: $u_x = u \cos \theta$

Vertical component: $u_y = u \sin \theta$

where:

u = launch speed (typically m s^{-1})

u_x = horizontal component of the launch velocity

u_y = vertical component of the launch velocity

θ = angle of the launch velocity relative to the horizontal

The vertical component is affected by Earth's **gravitational field**.

Neglecting air resistance, the horizontal velocity does not change and remains as $u \cos \theta$ for the entire time of flight.

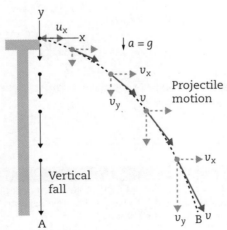

FIGURE 1.8 Two balls are released simultaneously from the same height, one with a horizontal component and the other without. They hit the ground at the same time.

9780170459174

1.2.2 Vertical velocity

The vertical component of motion is affected by the constant gravitational **field**, g.

It has a constant acceleration of -9.80 m s^{-2}.

Vertically $u_y = v_y$ only when $t = 0$. Horizontally $u_x = v_x$.

Using this known acceleration, the kinematics (or s, u, v, a, t) equations can be applied to analyse the vertical components.

$$v_y = gt + u_y$$

$$s_y = \frac{1}{2}gt^2 + u_y t$$

$$v_y^2 = u_y^2 + 2gs_y$$

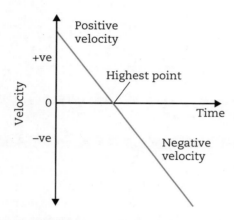

FIGURE 1.9 When drag is considered negligible, the vertical component of motion of a projectile proceeds at constant negative acceleration: $g = -9.80$ m s^{-2}.

1.2.3 Determining range

Hint

At a projection angle of 45° the range is a maximum. This can be deduced from the equations for projectile motion.

The range of a projectile depends on the angle of projection and the initial velocity. The range can be determined by using the equation $s_x = u_x t$.

Hint

Key formula

Maximum range formula for projectile motion

$$R = (v_0 \cos \theta) \times T_{\text{flight}}$$

$$R = (v_0 \cos \theta) \times \frac{2(v_0 \sin \theta)}{g}$$

The range is maximised at an angle of 45° travelling at a consistent horizontal velocity of $v_0 \cos \theta$

for the total time of flight of $\dfrac{2(v_0 \sin \theta)}{g}$

Hint

If the projectile lands on level ground, the maximum height of the object will occur at exactly half the time of its flight. The range of this projectile can be calculated by using double the time to maximum height (as the object is exactly halfway through its course). This only occurs when the parabolic path is symmetrical.

The maximum height can be calculated when the vertical velocity is 0 m s^{-1}.

1.3 Inclined planes

1.3.1 Forces on inclined plane

An inclined plane is a surface that has been tilted at an angle with respect to the horizontal. A mass on an inclined plane is subject to forces parallel and perpendicular to the plane. These forces can be combined and analysed in terms of vector sums.

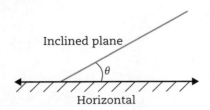

FIGURE 1.10 The angle of inclination, θ, is an angle relative to the horizontal.

On an inclined plane, there are multiple forces acting on the object, although not all problems take all forces into consideration. The forces include:
1   Weight force (that always acts down)
2   Normal force (that acts perpendicular to the surface)
3   Friction force (that opposes motion)
4   Applied force (that acts in a specific direction)

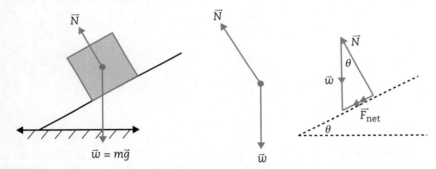

FIGURE 1.11 Inclined plane: for a frictionless surface, the net force is the vector sum of the normal force and the weight force.

The forces shown here are drawn in free body force diagrams. Free body force diagrams show the relevant forces acting on an object drawn as arrow heads from the centre of mass of the object proportional to their magnitude and in the direction that they act.

Hint

Key formula

Parallel to surface	Perpendicular to surface
$\Sigma F_{\parallel} = mg \sin \theta = ma$	$\Sigma F_{\perp} = N - mg \cos \theta = 0$
$\Rightarrow a = g \sin \theta$	where:
where:	$\Sigma F_{\perp}$ = net force perpendicular to the surface
$\Sigma F_{\parallel}$ = net force parallel to the surface (N)	N = normal force (N)
m = mass (kg)	m = mass (kg)
g = gravitational force = 9.8 m s^{-2}	g = gravitational force = 9.8 m s^{-2}
θ = angle of inclination (°)	θ = angle of inclination (°)
a = acceleration along the slope (m s^{-2})	

1.3.2 Weight force

Due to the **angle of inclination**, the **weight** force ($F_w = mg$) is not acting parallel to, or perpendicular to the surface. This means, the weight force must be split into its parallel and perpendicular **components**, which can be done using trigonometry.

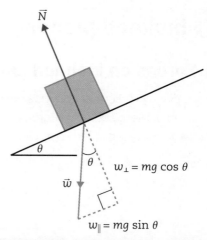

FIGURE 1.12 Perpendicular and parallel components of forces acting on a box on an incline. The weight force may be split into its parallel and perpendicular components.

Hint
Key formula
Perpendicular or y-component = $mg \cos \theta$
Parallel or x-component = $mg \sin \theta$
The greater the angle θ, the greater the weight force acting down the plane.

1.3.3 Normal force

Normal force = weight force (perpendicular component)
   This is because the net force perpendicular to the surface must equal zero as the object is not accelerating up or down. Therefore the normal force upward will always be equal to the force downward.

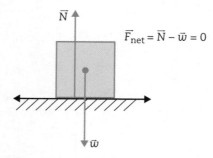

FIGURE 1.13 Horizontal plane: the net force acting on a stationary box is zero. In the vertical direction the sum of the forces is $\Sigma F = N - w = 0$. The normal force is equal in magnitude and opposite in direction to the weight force.

1.3.4 Friction force

When friction is involved, the net force parallel to the surface has an additional term:
   $F_{net} = mg \sin \theta - F_{friction} = ma$
   If the object is not accelerating, then $mg \sin \theta = F_{friction}$

Hint
Key formula
$F_{net} = mg \sin \theta - F_{friction} = ma$.

1.4 Circular motion

1.4.1 Uniform circular motion

Uniform circular motion is the motion of an object in a circle at a constant speed. The uniform term reflects that the speed (not velocity) is constant. Any object travelling in circular motion has a net force acting towards the centre of the circle; that is, it is constantly accelerating. The velocity is constantly changing too, due to its changing direction. An object circling around a point must be continuously pushed inwards. Therefore, the force acting upon the object is constantly perpendicular (inward) to the velocity of the object. The acceleration is always directed towards the exact centre of the motion. The pointing to the centre is called **centripetal**, which means centre-seeking.

Hint
Key formula
$$v = \frac{2\pi r}{T} = 2\pi f$$
where:
r = radius (m)
T = period (s)
f = **frequency** (Hz)
v = average and instantaneous speed (m s^{-1})

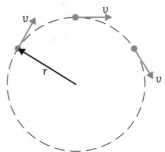

FIGURE 1.14 In circular motion, the velocity vector changes direction continuously. It travels the circumference of the circle in one time period.

1.4.2 Speed and period

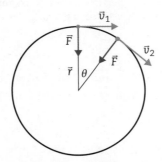

FIGURE 1.15 Derivation of centripetal acceleration: two equal magnitude velocity vectors are shown separated by a time interval, which is represented by the angle between the equal magnitude radius vectors.

Average speed (Note that speed is not velocity – it does not have a direction)

$$\text{Speed} = \frac{\text{distance}}{\text{time}}$$

Distance for one period = circumference = $2\pi r$

Hence, speed = $\dfrac{\text{circumference}}{\text{period}}$, $\mathbf{v} = \dfrac{2\pi \mathbf{r}}{\mathbf{T}}$

Objects undergoing uniform circular motion have a constant and unchanging speed. The **period** is the time it takes to complete one revolution. The related formulas to solve problems in circular motion include:

$$a_c = \frac{v^2}{r}$$

$$v = \frac{2\pi r}{T}$$

$$F = \frac{mv^2}{r}$$

1.4.3 Centripetal acceleration and force

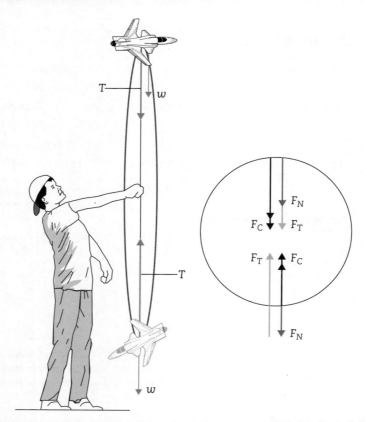

FIGURE 1.16 An object being whirled in a vertical circle. Note the force centripetal is constant to maintain a circular motion. The force centripetal is the sum of the force weight and the force tension in the string.

Centripetal force

The **centripetal force** is the sum of all forces that points toward the centre of the circle. It is the net force. It is not a type of force, rather a summation of all real forces.

Centripetal acceleration

The centripetal acceleration is directed radially toward the centre of the circle and has a magnitude equal to the square of the body's speed along the curve divided by the radius $a_c = \dfrac{v^2}{r}$

> **Hint**
>
> **Key formula**
>
> $F = ma$
>
> $a = \dfrac{v^2}{r}$
>
> Therefore: $F = \dfrac{mv^2}{r}$

Forces that may typically provide a centripetal acceleration include electrostatic forces, force of tension, gravitational force (weight) and electromagnetic forces.

1.5 Gravitational force and fields

1.5.1 Newton's law of universal gravitation

Newton's law of universal gravitation states that every point mass in the universe attracts every other point mass with a force that is directly proportional to the product of their masses and inversely proportional to the square of the distance between them ($F_{net} = \dfrac{GMm}{r^2}$).

It is important to note the equation requires the distance between the two centres of mass of the objects. This is significant when the equation is used to determine the force a planet exerts on a **satellite**, as the radius of the planet may have to be added if the question provides only the altitude.

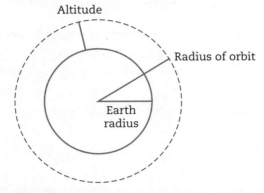

FIGURE 1.17 The orbit radius of a satellite may be provided or may require the addition of the satellite's altitude and the radius of the planet. ($R_{orbit} = R_{planet}$ + altitude).

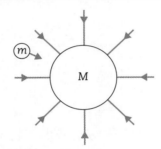

FIGURE 1.18 The gravitational field surrounding a mass mediates the gravitational force, accelerating the objects towards each other.

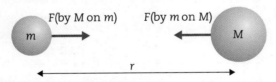

FIGURE 1.19 Gravitational force acts between bodies with mass with an equal and opposite force, in accordance with Newton's law of universal gravitation.

Hint

Gravitational potential energy is *not* stored in the object; rather, it is stored in the field.

1.5.2 Gravitational potential energy

Gravitational potential energy is due to the interaction of objects via their gravitational fields. It is the gravitational field that mediates or exerts the force on one object due to the mass of another. Hence, we say that the gravitational field does **work** when an object falls in Earth's gravitational field. In this case, the work is done by the field on the object because the object moves in the direction of the field. The kinetic energy increases and the **potential energy** decreases.

Hint

Remember that potential energy belongs to a system of interacting objects. It is not meaningful to refer to the potential energy of a single object.

1.5.3 Gravitational field strength

Each fundamental force can be described as acting by the means of a field.

The gravitational field is a model used to explain the influence that a massive body (body with mass) extends into the space around itself, thus producing a force on another massive body.

The gravitational field model allows us to:

- Explain how objects can exert forces without being in contact
- Predict the acceleration of an object
- Calculate the mass of an object from the observed forced exerted on/by another object.
- Calculate the mass of distant objects by observing their orbits.

The gravitational field strength can be calculated using the known equations $F_w = mg$ and $F_{net} = \dfrac{GMm}{r^2}$.

$$g = \frac{F}{m} = \frac{GM\cancel{m}}{\cancel{m}r^2}$$

$$\text{Hence, } g = \frac{GM}{r^2}$$

On Earth, the gravitational field strength is 9.80 m s^{-2}.

Since $g = \dfrac{F}{m}$, it can be noted that the gravitational field asscociated with M exists independent of the secondary mass in the field, that is, g is the same for all objects on Earth, regardless of mass.

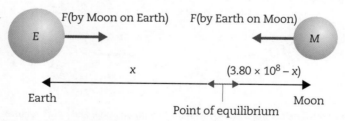

FIGURE 1.20 Diagram of the Earth–Moon system. There is a point of gravitational equilibrium that exists between Earth and the Moon where the forces are equal in magnitude and opposite in their direction.

1.6 Orbits and Kepler's laws of planetary motion

1.6.1 Kepler's first law: the law of ellipses

All planets move in elliptical **orbits** with the Sun at one focus.

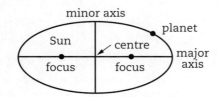

FIGURE 1.21 Kepler's first law, the law of **ellipses**, illustrates that the path of the planets about the Sun is elliptical in shape, the Sun being at one focus.

> **Hint**
> To recall this law, think – the Sun comes first. Therefore the first law is the one that has the Sun at the centre or at one focus.

1.6.2 Kepler's second law: the law of equal areas

The velocity of planets changes throughout their orbit. The closer the planets are to the Sun, the greater their speed. Although the speed changes, Kepler concluded that the areas swept out in equal time intervals are the same.

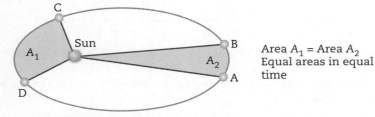

FIGURE 1.22 In Kepler's second law, segments AB and CD are swept out in equal time intervals (Kepler's second law).

A line that connects a planet to the Sun sweeps out equal areas in equal time periods.

> **Hint**
> To recall this law, think – the area is the second in the sequence of perimeter, area, volume. Therefore the law of equal areas is the second law.

Hint
To recall this law, think that R is to the power of 3, therefore it is the third law.

1.6.3 Kepler's third law: the law of periods

The square of the period of a planet's orbit is proportional to the cube of its **mean orbital radius** ($T^2 \propto R^3$). This is evident in the formula.

$$\frac{T^2}{R^3} = k = \frac{4\pi^2}{GM}$$

The period squared over the separation cubed is equal to a constant, k, for all objects orbiting the same central body. For example, all planets circling the Sun have a constant k value. This formula is true for any orbiting system, but the constant, k, may differ between systems.

For example, the many moons that circle Jupiter will have the same constant k value but it will be different from the k value for the planets circling the Sun.

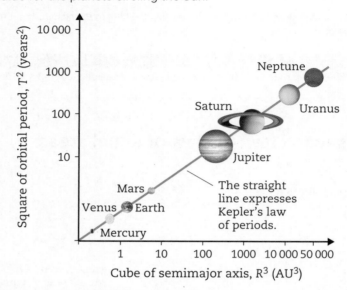

FIGURE 1.23 Kepler's third law, the law of periods, is constant for all orbiting bodies within a given system. This is illustrated for the planets of the solar system.

Deriving Kepler's third law

Newton's law of universal gravitation and uniform circular motion can be used to quantify the relationship predicted in Kepler's third law.

$$F_c = \frac{mv^2}{R}, \quad F_g = \frac{GMm}{R^2}$$

$$\frac{\cancel{m}v^2}{\cancel{R}} = \frac{GM\cancel{m}}{R^{\cancel{2}}}$$

$$\text{But } v^2 = \frac{(2\pi R)^2}{T^2} = \frac{4\pi^2 R^2}{T^2}$$

$$\text{So } \frac{4\pi^2 R^2}{T^2} = \frac{GM}{R}$$

$$\Rightarrow \frac{4\pi^2}{GM} = \frac{T^2}{R^3}$$

Therefore,

$$\frac{T^2}{R^3} = k = \frac{4\pi^2}{GM}$$

Glossary

angle of inclination
angle, θ, relative to the horizontal; $0° < \theta < 90°$

centripetal
centre-seeking; directed towards the centre

centripetal force
in uniform circular motion, the sum of real forces that points towards the centre of the circle

components
two or more vectors into which a vector can be resolved

ellipse
a regular, curved shape that is a conic section (formed by cutting a cone obliquely); the path of satellites in orbit around larger bodies

field
the means by which action-at-a-distance forces are exerted

frequency
number of times a circular motion is completed in a time period

gravitational field
the field that mediates the gravitational force between all objects with mass; the field surrounding all objects with mass: $g = \dfrac{GM}{r^2}$

gravitational potential energy
the potential energy associated with the interaction of objects via the gravitational force; the potential energy is stored in the gravitational field

mean orbital radius
the average radius of orbit of one massive object about another, e.g. of Earth revolving about the Sun

normal force
the support force exerted upon an object that is in contact with another stable object

orbit
a regularly repeated elliptical path of one object about another massive object, such as a planet about a star

A+ DIGITAL FLASHCARDS
Revise this topic's key terms and concepts by scanning the QR code or typing the URL into your browser.

https://get.ga/aplus-qce-phys-u34

period
time taken for an object undergoing circular motion to complete one revolution

potential energy
energy stored in a system due to the interaction of components in the system via forces; energy stored in a field. Potential energy gives a system the ability to do work

rectangular components
components that are at right-angles to each other; perpendicular components

resolute
component of a vector

satellite
a natural (e.g. moon) or synthetic (e.g. GPS or communications satellite) body that orbits a significantly larger mass

scalar multiplier
positive or negative number that can change the magnitude and/or reverse the direction of a vector

weight
the gravitational force that acts on an object, $F_w = mg = \dfrac{GMm}{r^2}$

work, W
energy transferred due to the action of a force: $W = Fs$

U3 – TOPIC 1 – GLOSSARY

Revision summary

Use the following summary of syllabus dot points and key knowledge within Unit 3 Topic 1 to ensure that you have thoroughly reviewed the content. Provide a brief definition or comment for each item to demonstrate your understanding or code them using the traffic light system – Green (all good); Amber (needs some review); Red (priority area to review). Alternatively, write a follow up strategy.

The cognitive verbs have been identified in bold.

Gravity and motion	
Vectors	
• **use** vector analysis to resolve a vector into two perpendicular components	
• **solve** vector problems by resolving vectors into components, adding or subtracting the components and recombining them to determine the resultant vector	
Projectile motion	
• **recall** that the horizontal and vertical components of a velocity vector are independent of each other	
• **apply** vector analysis to determine horizontal and vertical components of projectile motion	
• **solve** problems involving projectile motion	
• Mandatory practical: **Conduct** an experiment to determine the horizontal distance travelled by an object projected at various angles from the horizontal	
Inclined planes	
• **solve** problems involving force due to gravity (weight) and mass using the mathematical relationship between them	
• **define** the term *normal force*	
• **describe** and represent the forces acting on an object on an inclined plane through the use of free-body diagrams	
• **calculate** the net force acting on an object on an inclined plane through vector analysis	››

››	**Circular motion**	
	• **describe** uniform circular motion in terms of a force acting on an object in a perpendicular direction to the velocity of the object	
	• **define** the concepts of *average speed* and *period*	
	• **solve** problems involving average speed of objects undergoing uniform circular motion	
	• **define** the terms *centripetal acceleration* and *centripetal force*	
	• **solve** problems involving forces acting on objects in uniform circular motion	
	Gravitational force and fields	
	• **recall** Newton's law of universal gravitation	
	• **solve** problems involving the magnitude of the gravitational force between two masses	
	• **define** the term *gravitational fields*	
	• **solve** problems involving the gravitational field strength at a distance from an object	
	Orbits	
	• **recall** Kepler's laws of planetary motion	
	• **solve** problems involving Kepler's third law	
	• **recall** that Kepler's third law can be derived from the relationship between Newton's law of universal gravitation and uniform circular motion	

© State of Queensland (QCAA) Physics General Senior Syllabus 2019 CC BY 4.0 https://creativecommons.org/licenses/by/4.0/

Exam practice

Multiple-choice questions

Each multiple-choice question is worth 1 mark.

Solutions start on page 120 .

Question 1

> **Hint**
>
> When answering a question asking you to *determine*, remember to follow the QCAA definition.
>
> **determine**: establish, conclude or ascertain after consideration, observation, investigation or calculation; decide or come to a resolution.

Determine which of the following definitions describes the term 'normal force' most accurately.

A the component of a contact force parallel to the surface that an object is in contact with

B the force acting along an imaginary line drawn perpendicular to the surface

C the force applied to an object by a person or another object

D the resistance to motion of an object moving relative to another object

Question 2

A ball is thrown with an initial velocity of $25\,\mathrm{m\,s^{-1}}$ at an angle of $15°$ to the horizontal. The horizontal and vertical components of the velocity, respectively, are:

A $6.34\,\mathrm{m\,s^{-1}}$; $13.6\,\mathrm{m\,s^{-1}}$

B $6.47\,\mathrm{m\,s^{-1}}$; $24.2\,\mathrm{m\,s^{-1}}$

C $13.6\,\mathrm{m\,s^{-1}}$; $6.34\,\mathrm{m\,s^{-1}}$

D $24.2\,\mathrm{m\,s^{-1}}$; $6.47\,\mathrm{m\,s^{-1}}$

Question 3

> **Hint**
>
> When answering a question asking you to *select*, remember to follow the QCAA definition.
>
> **select**: choose in preference to another or others; pick out

> **Hint**
>
> When answering a question asking you to consider a *relationship*, identify the appropriate formula that relates the two variables, in this case, $F_c = \dfrac{mc^2}{r}$ and write a proportionality statement, e.g. $F_c \propto v^2$.

Select the diagram that best represents the mathematical relationship between centripetal force and velocity for an object undergoing uniform circular motion.

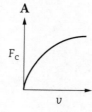 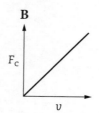 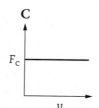 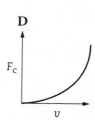

Question 4

An object projected horizontally out of a window $12\,\mathrm{m}$ off the ground lands $4.2\,\mathrm{m}$ out from the base of the building. The velocity at which it was projected horizontally may be determined to be:

A $2.68\,\mathrm{m\,s^{-1}}$

B $3.78\,\mathrm{m\,s^{-1}}$

C $4.53\,\mathrm{m\,s^{-1}}$

D $7.71\,\mathrm{m\,s^{-1}}$

Question 5

A long play (LP) record player has a maximum radius of 21 centimetres and rotates at 33.3 revolutions per minute (rpm). The average speed, in m s^{-1}, for a point on the outer rim of the record is:

A 0.04 m s^{-1}

C 0.73 m s^{-1}

B 0.56 m s^{-1}

D 2.38 m s^{-1}

Question 6

Planet X follows an elliptical orbit about a star and is known to have an average velocity of 22 km s^{-1} when at its average distance from the star. Select the response that is most likely to be the planet's velocity when it is furthest from the star.

A 15 km s^{-1}

C 30 km s^{-1}

B 22 km s^{-1}

D 32 km s^{-1}

Question 7

An object is swung in an anticlockwise horizontal circle at a constant speed. Select the diagram that illustrates the direction of the velocity vector and acceleration vector at point X.

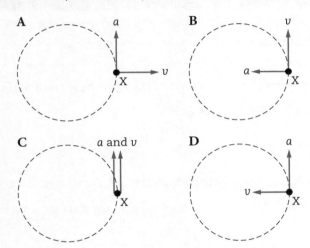

Question 8

The diagram below shows the velocity of a projectile in the vertical direction over time.

Determine which of the following is the best description of the initial motion of the projectile.

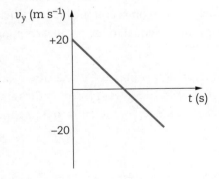

A dropped initially from rest

C thrown initially vertically down

B projected horizontally

D thrown initially vertically up

Question 9

A person normally weighs 550 N on the surface of Earth. Determine their weight on the surface of the Moon. Use the mass and radius of the Moon as 7.35×10^{22} kg and 1.74×10^3 km, respectively.

A 34 kg

C 56 kg

B 55 kg

D 91 kg

Question 10

The mean distance of Mercury from the Sun is 57.91 million km, while the distance of Earth from the Sun is 150 million km. The period of revolution of Mercury around the Sun can be determined as:

A 0.064 years

C 1.23 years

B 0.24 years

D 1.32 years

Short response questions

Question 11 (2 marks)

> **Hint**
>
> When answering a question asking you to *determine*, remember to follow the QCAA definition.
>
> **determine**: establish, conclude or ascertain after consideration, observation, investigation or calculation; decide or come to a resolution.

Kepler's third law, the law of periods, may be used to determine the mass of the Sun.

This law is often expressed mathematically as below.

$$\frac{T^2}{r^3} = \frac{4\pi^2}{GM}$$

Determine the mass of the Sun using Kepler's third law and the measured facts below.

Earth is 1 AU from the Sun. 1 AU is the astronomical unit, the distance from Earth to the Sun, with a value of 150 million km.

Mercury is 0.387 AU from the Sun.

The mass of Earth is 5.98×10^{24} kg. Earth orbits the Sun with a period of 1 year.

Question 12 (4 marks)

Students were asked to investigate the relationship between the angle of elevation of a frictionless ramp and the parallel component of the weight of an object on the incline. Their aim was to test a hypothesis that the weight of the hanging mass would be equal to the parallel component of the trolley cart's weight down the incline for each angle.

An experiment was set up with a 1.2 m long ramp and its elevation was varied through a range of angles from 10° to 50°. A trolley cart of mass 75.5 g was placed on the ramp and a hanging mass was placed so as to hang freely over a pulley at the top of the ramp, holding the trolley cart stationary.

Table 1.1 shows the results of the experiment.

Angle, θ (°)	Hanging mass (kg)
10.0	0.131
20.0	0.263
30.0	0.379
40.0	0.488
50.0	0.588

TABLE 1.1 Inclined plane experimental values

a   Draw a free body diagram of the trolley cart on the incline at 40°, including labelled values
    for all calculated forces acting.   2 marks

b   Extend Table 1.1 with the further calculated values for $\sin\theta$ and force weight parallel to
    the slope to then draw a conclusion based on the aim and hypothesis.   2 marks

> **Hint**
>
> When answering a question asking you to *draw conclusions*, remember to follow the QCAA definition.
>
> **draw conclusions**: make a judgment based on reasoning and evidence

Angle, θ (°)	Hanging mass (kg)		
10.0	0.131		
20.0	0.263		
30.0	0.379		
40.0	0.488		
50.0	0.588		

Question 13 (4 marks)

> **Hint**
>
> When answering a question asking you to *calculate*, remember to follow the QCAA definition.
>
> **calculate**: determine or find (e.g. a number, answer) by using mathematical processes; obtain a numerical answer showing the relevant stages in the working; ascertain/determine from given facts, figures or information

a   Calculate the gravitational force acting on a satellite of orbital radius 5500 km and
    mass 2000 kg. Use mass_{Earth} as 5.98×10^{24} kg and the radius of Earth as 6.37×10^6 m.   2 marks

b   Determine the orbital period, in hours, of a satellite of mass 750 kg and altitude of
    8000 km. Use mass_{Earth} as 5.98×10^{24} kg and the radius of Earth as 6.37×10^6 m.   2 marks

Question 14 (2 marks)

A ball strikes the ground with a resultant velocity of 150 m s^{-1}. It is known that the horizontal component
of the velocity is 62 m s^{-1} to the right. Determine the vertical component of the final velocity and the angle
of impact relative to the ground.

Question 15 (4 marks)

An 80 kg crate is pulled up an inclined plane that makes an angle of 10° with the horizontal. The force of tension, T, applied to the rope makes an angle of 15° with the inclined plane.

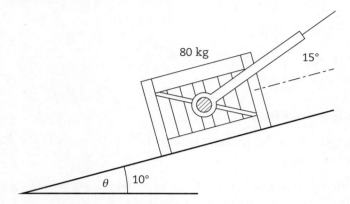

The crate is measured to accelerate up the plane with an acceleration of $a = 1.0 \, \mathrm{m \, s^{-2}}$.

Determine the force of tension, T, applied to the rope. Show all working required.

Question 16 (5 marks)

A 25 kg mass has three forces applied to it, as shown below.

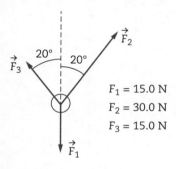

$F_1 = 15.0 \, \mathrm{N}$
$F_2 = 30.0 \, \mathrm{N}$
$F_3 = 15.0 \, \mathrm{N}$

Determine the net force applied to the mass and its resulting acceleration.

Question 17 (8 marks)

A mass is thrown at an angle of 45° and achieves a range of 96.90 m on horizontal ground. Calculate the following values.

a the total time of flight.   3 marks

b the maximum height reached above the ground.   3 marks

c the initial velocity.   1 mark

d the impact velocity of the mass at landing.   1 mark

QUESTION 18 (6 marks)

The figure below shows a 200 kg satellite orbiting Earth as part of the Global Positioning System (GPS) network. Each satellite transmits signals in the radio wavelengths to various receivers on Earth.

When the satellite is directly overhead, a signal from the satellite to a receiver on the surface of the Earth is measured to take a period of 0.065 seconds. Radio waves, like all light, travel at the speed of light, c (3×10^8 m s^{-1}). The radius of Earth is 6.37×10^6 m.

Note: additional values required may be found in QCAA *Physics Formula and Data book*.

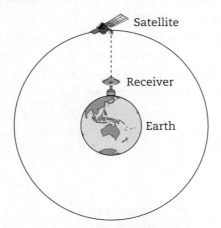

a Calculate the height of the satellite above the surface of Earth. 1 mark

b Determine the gravitational force acting on the satellite that keeps it in orbit. 3 marks

c Explain why satellites may maintain a constant speed while still accelerating towards the centre of Earth. 2 marks

Question 19 (6 marks)

Hint

When answering a question asking you to *recall*, remember to follow the QCAA definition.

recall: remember; present remembered ideas, facts or experiences; bring something back into thought, attention or into one's mind

a Recall Kepler's three laws of planetary motion. 3 marks

b The orbital period for three consecutive months was measured for an artificial satellite to monitor its orbit (see Table 1.2). Calculate the average orbital period and the mean altitude over this three-month period. Show your working. 2 marks

Orbital data	Month 1	Month 2	Month 3
Orbital period (minutes)	96.2	95.4	91.0
Mean altitude (km)	219	216	210

TABLE 1.2 Orbital data from an Earth satellite

c Calculate approximately how many orbits of Earth the satellite makes within a three-month (90-day) period. 1 mark

Question 20 (3 marks)

Determine the gravitational force of attraction between an electron and an atom's nucleus that are a distance of 1.5 nm apart. The mass of the electron is 9.11×10^{-31} kg and you are to consider the nucleus to be made up of a proton and a neutron, each of mass 1.67×10^{-27} kg.

Chapter 2
Topic 2: Electromagnetism

Topic summary

There are four fundamental forces. The strong force, the weak force, the gravitational force and the electromagnetic force. The electromagnetic force is a combination of both the electrostatic and magnetic forces. Electromagnetism introduces students to concepts in electrostatics, including Coulomb's law, electric fields and electric field strength, and concepts in magnetism, including magnetic fields and the forces on charged particles moving within magnetic fields. Electromagnetic induction is demonstrated through Faraday's law of induction and Lenz's law, applying these concepts to the production and transmission of alternating current. The impact of the discovery of electromagnetic induction cannot be understated as it resulted in the development of the electric generator and motor. Electromagnetic theory unifies the phenomena of electricity and magnetism with field theory and electromagnetic equations. Maxwell also described many of the properties of light and predicted that light was an electromagnetic wave.

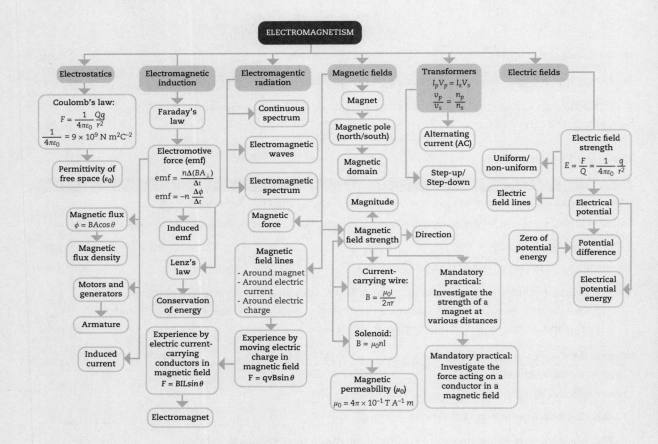

2.1 Electrostatics

$\vec{F}$(by q on Q) $\vec{F}$(by Q on q)

Q q

FIGURE 2.1 An example of two like-charged particles. They exert the exact same repulsive force on each other. This causes the particle Q to move to the left (from the force exerted by charge q), and causes the charge q to move to the right (from the force exerted by charge Q).

2.1.1 Coloumb's law

Coulomb's law is the second law of electrostatics. It states that the force between two point charges is:

- directly proportional to the product of their electric charges
- inversely proportional to the square of the distance between them
- inversely proportional to the permittivity of the surrounding medium.

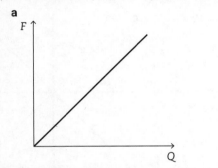

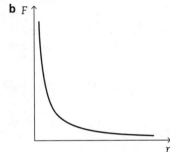

FIGURE 2.2 **a** A linear function. This function has a constant gradient, showing that the factor by which Q changes will directly affect the force F in the same fashion. **b** An inverse-squared function. This function shows that as the separation distance is increased, the force F between the two charged particles decreases dramatically.

Worked example

A charge q = +2.0 nC is separated from a second charge Q = -4.0 nC by a distance of r = 0.002 m.

Determine the electrostatic force acting between the two charges.

Answer

1 State the equation.

$$F = \frac{kqQ}{r^2}$$

2 Substitute in the values, being careful to convert to SI units and account for metric prefixes.

$$F = \frac{kqQ}{r^2}$$

$$F = \frac{9.0 \times 10^9 \times +2 \times 10^{-9}\,C \times -4 \times 10^{-9}}{(0.002)^2}$$

$$F = 1.8 \times 10^{-2}\,N$$

> **Hint**
>
> **Key formula**
>
> **Coulomb's law**
>
> $$F = \frac{1}{4\pi\varepsilon_o}\frac{qQ}{r^2}$$
>
> where:
>
> F = electrostatic force q exerts on Q (N)
>
> q = charge (C) of one point charge
>
> Q = charge (C) of the other point charge
>
> r = separation distance of q and Q (m)
>
> ε_0 = **electrical permittivity** of free space
>
> $\dfrac{1}{4\pi\varepsilon_o} \approx 9.0 \times 10^9$ Nm^{2}C^{-2}; also known as Coulomb's constant, k
>
> Coulomb's law (simplified)
>
> $$F = \frac{kqQ}{r^2}$$

2.1.2 Electric fields

All objects with electric charge emanate an **electric field**. The strength of the **field** is dependent upon:
- the size of the charge
- the distance from the charged object

An electric field is a region around an electrically charged particle or object within which a force would be exerted on other electrically charged particles or objects.

Electric field lines

Electric field lines show the direction of the force on a small positively-charged test particle, hence field lines are drawn radially outward (perpendicular to the surface) from a positive charge and are drawn radially inward toward a negative charge.

Field lines never cross – they represent the net lines of force.

The field strength is proportional to the density of the field lines - the more dense the field lines, the greater the field strength.

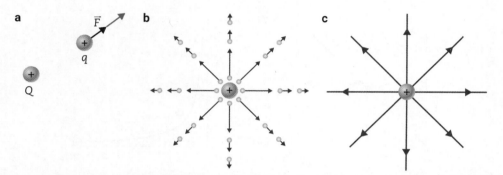

FIGURE 2.3 a A test charge is accelerated away from the positive source charge. **b** The size of the vector decreases as the distance between test charge and source charge increases. **c** Join up vectors to obtain electric field lines. Note that field lines always point *away* from a positive charge.

Electric field strength

The electric field strength is a measure of the intensity of an electric field at a particular point. It is defined as the force per unit charge that acts on a small positive test charge at that point.

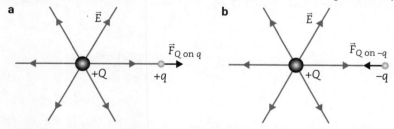

FIGURE 2.4 a A large positive charge repels a small positive test charge. **b** A large positive charge attracts a small negative test charge.

The electric field strength at any point can be determined using:

$$E = \frac{F}{q}$$

or

$$E = \frac{1}{4\pi\varepsilon_0}\frac{Q}{r^2}$$

Hint

Key formula

Electric field strength

$$E = \frac{F}{q}$$

where:

E = electric field strength (N C^{-1})

F = force acting on test charge q (N)

q = charge of the test object in the field (C)

Hint

Key formula

Electric field strength when q is unknown

$$E = \frac{1}{4\pi\varepsilon_0}\frac{Q}{r^2}$$

where:

Q = large charge (C)

r = distance in metres (m) between q and Q

$\frac{1}{4\pi\varepsilon_0} \approx 9 \times 10^9$ Nm^{2}C^2; also known as Coulomb's constant, k

Electric field strength simplified:

$$E = \frac{kQ}{r^2}$$

2.1.3 Electrical potential energy

Just like an object in a gravitational field has gravitational potential energy, an electric charge in an electric field has **electrical potential energy**. While gravitational fields do **work** on falling objects, electric fields do work on charged objects.

The **electrical potential** is defined as the potential energy per unit charge, $V = \frac{U}{q}$, measured as joules per coulomb, or volts.

When considering a charge in an electric field, it is important to consider the electrical potential.

The difference in potential energy for a charge at difference points is called the **potential difference**, or more commonly, voltage. It is the work done per unit charge .

Potential difference

Consider a negatively charged object and its movement towards another charge.

Moving toward a positively charged object: kinetic energy increases, thus the electrical potential decreases. There is a negative potential difference.

Moving toward a negatively charged object: kinetic energy decreases, thus the electrical potential increases. There is a positive potential difference.

Consider a positively charged object and its movement towards another charge.

Moving toward a positively charged object: kinetic energy decreases, thus the electrical potential increases. There is a positive potential difference (or potential rise).

Moving toward a negatively charged object: kinetic energy increases, thus the electrical potential decreases. There is a negative potential difference (or potential drop).

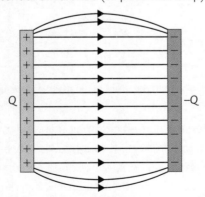

FIGURE 2.5 Uniform electric field formed between the plates of a parallel plate capacitor. The field is uniform in the section between the plates.

> **Hint**
>
> **Key formula**
>
> Electrical potential
>
> $$V = \frac{U}{q}$$
>
> where:
>
> V = electrical potential of a charge (V)
>
> U = **potential energy** (J)
>
> q = magnitude of the charge in the field (C)
>
> **Key formula**
>
> Potential difference
>
> $$\Delta V = \frac{\Delta U}{q}$$
>
> where:
>
> ΔV = potential difference between two points (V)
>
> ΔU = change in potential energy (J), also known as the work done on the charge
>
> q = magnitude of the charge moving in the field (C)

2.2 Magnetic fields

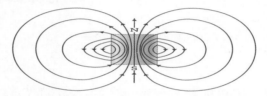

FIGURE 2.6 Magnetic field lines coming out of a bar magnetic. A bar **magnet** is a bar of magnetically aligned metal that has its domains in a similar direction. Note that the arrows point *out* from the **north pole**, and *in* to the **south pole** outside of the magnet.

2.2.1 Magnetic domains

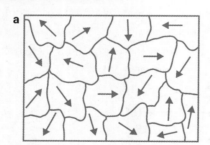 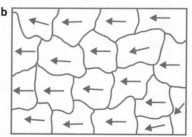

FIGURE 2.7 a Randomly aligned domains. The material is not magnetised as a result. **b** Domains are lined up. This material would be magnetised.

A **magnetic field** is the field created by magnetic domains and by moving electric charges, including magnetic materials and electric charges moving through a wire (ie. electric current).

If these internal **magnetic domains** are generally aligned, the object is deemed to be magnetic. The strength of the magnetic field is dependent on the proportion of aligned internal domains.

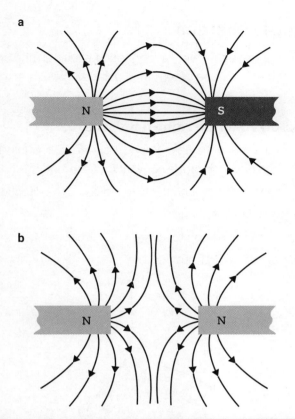

FIGURE 2.8 a Two unlike poles (north and south) brought together attract each other due to their field lines lining up. **b** Two like poles brought together repel each other.

2.2.2 Representing a magnetic field

If a material is magnetised, it will have a north pole and south pole.

Magnetic fields are drawn coming out from the north pole and coming in at the south pole.

Opposite poles attract. Similar poles repel.

The density of field lines reflects the strength of the magntic field.

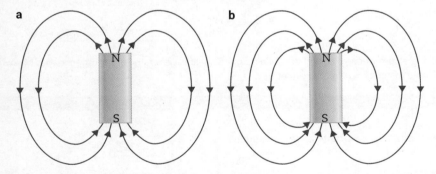

FIGURE 2.9 Magnetic field lines represent field strength by their density. **a** A weaker magnetic field **b** A comparatively stronger magnetic field.

2.2.3 Moving electric charges

If an electric current is flowing through a wire, a magnetic field is formed around the wire.
The direction of the magnetic field can be determined using Maxwell's screw rule
(the right-hand rule).

Applying Maxwell's screw rule (the right-hand rule)

To determine the direction of the magnetic field around a current-carrying wire:

- Point your right thumb in the direction of the conventional current
- The direction that your fingers wrap around the wire indicates the direction of the magnetic field.
- The strength of the magnetic field depends upon:
  - The magnitude of the current through the wire
  - The distance of the magnetic field from the wire.

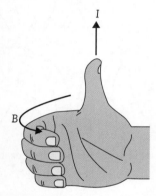

FIGURE 2.10 Maxwell's screw rule shows the direction of the magnetic field around a current-carrying conductor. Point your right thumb in the direction of conventional current, and your fingers will wrap around in the direction of the magnetic field.

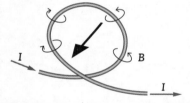

FIGURE 2.11 Outside a current-carrying coil of wire a magnetic field is generated in accordance with Maxwell's screw rule. This allows a solenoid to act as an electromagnet.

Hint

Key formula

Magnetic field strength from a current-carrying conductor

$$B = \frac{\mu_0 I}{2\pi r}$$

where:

B = magnetic field strength distance r from the current-carrying conductor (T)

μ_0 = **magnetic permeability** of free space ($4\pi \times 10^{-7}$ T m A^{-1})

I = current travelling through the conductor (A)

r = perpendicular distance from the current-carrying conductor (m)

2.2.4 Solenoids

Solenoids are coils of current-carrying wire that are used to create a large and approximately uniform field within the coil.

Each loop of wire creates a magnetic field. The sum of these fields within the coil add up to give a large and approximately uniform magnetic field, making this device an **electromagnet**.

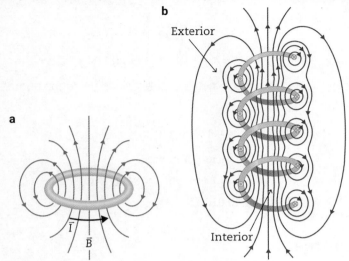

FIGURE 2.12 a Magnetic field due to a single loop of wire. **b** Magnetic field due to a solenoid. Note that the field lines resemble those of a bar magnet.

Reviewing the variables within the formula $B = \mu_0 nI$ indicates that increasing the number of loops, n, will result in an increase in the strength of the magnetic field, B. Likewise, increasing the current, I, will also increase the strength of the magnetic field, B.

$$B \propto n$$

$$B \propto I$$

> **Hint**
>
> When performing calculations for solenoids, the number of turns, n, is a number per metre length, hence the actual number of turns, N, needs to be divided by the length of the solenoid, L, in metres.
>
> $$n = \frac{N}{L} \text{ turns per metre}$$

As is visible in Figure 2.12 above, the magnetic field lines are more spread out further away from the current-carrying conductor, and hence the magnetic field is weaker at an increased distance from the solenoid. Internally, though, the magnetic field is strong and uniform.

> **Hint**
>
> **Key formula**
>
> Magnetic field produced by a solenoid
>
> $$B = \mu_0 nI$$
>
> where:
>
> B = strength of the field inside the solenoid (T)
>
> μ_0 = permeability of free space ($4\pi \times 10^{-7}$ T m A^{-1})
>
> n = number of turns in the solenoid per metre (m^{-1})
>
> I = current travelling through the solenoid (A)

The direction of the magnetic field can be found through examining a single loop of wire and applying Maxwell's screw rule.

Worked example

A 10 cm long solenoid of diameter 2.5 cm and 800 turns of conducting wire has a current of 1.0 amp running through it. Determine the magnetic field strength, B, within the solenoid.

Answer

1   State the formula for the magnetic field inside a solenoid, $B = \mu_0 nI$.

2   Substitute the known values into the equation.

$$B = \mu_0 nI$$
$$B = 4\pi \times 10^{-7} \times n \times 1.0$$

Note here that n, the number of turns per metre, is determined using the number of turns and total length of the solenoid,

$$n = \frac{N}{L}, \text{ hence } n = \frac{800 \text{ turns}}{0.10 \text{ metres}}$$
$$n = 8000 \text{ turns per metre}$$

So now, $B = 4\pi \times 10^{-7} \times 8000 \times 1.0$

And $B = 0.010$ T

2.2.5 Magnetic force

The strength of the **magnetic force**, F, can be determined using:

$F = qvB \sin \theta$ for a single point charge

$F = BIL \sin \theta$ for a current-carrying conductor

The direction of the magnetic force can be determined using the right-hand rule.

Hint

Key formula

$F = qvB \sin \theta$ for a single point charge

$F = BIL \sin \theta$ for a current-carrying conductor

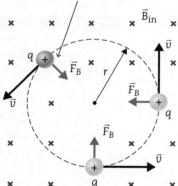

The magnetic force $\vec{F}_B$ acting on the charge is always directed towards the centre of the circle.

FIGURE 2.13 The magnetic force acting on a point charge is always directed towards the centre of a circular path, in accordance with the right-hand rule.

A current is a collection of charges moving in the same direction. A current experiences a force in a magnetic field, so we can expect that a single moving charge will also experience a force. When a force is exerted onto a moving charge by a magnetic field, the force depends on four factors:

1   The magnitude of the charge moving in the magnetic field, q

2   The magnitude of the velocity of the moving charge, v

3   The strength of the external magnetic field, B

4   The angle between the vectors v and B, θ

The magnitude of the magnetic force that a charged particle experiences is written mathematically as $F = qvB \sin \theta$.

> **Hint**
>
> **Key formula**
>
> Force on a moving charged particle
>
> $F = qvB \sin \theta$
>
> F = magnitude of the force on the moving charge (N)
>
> q = magnitude of the charge moving in the magnetic field (C)
>
> v = magnitude of the velocity of the moving charge (m s^{-1})
>
> B = strength of the external magnetic field (T)
>
> θ = angle between the vectors v and B (°)

> **Hint**
>
> If the particle enters the field with velocity perpendicular to the B field, it will experience the maximum magnetic force. This will cause the charged particle to follow a circular path, as per the right-hand rule. In this case, the centre-seeking force that the particle experiences causes it to undergo circular motion.

The magnetic force is the force a magnetic field exerts on a moving charge or current.

When a force is exerted onto a current by an external magnetic field, the force depends on four factors:

1   The magnitude of the current carried, I
2   The strength of the external magnetic field, B
3   The angle between the direction of the current and the field, θ
4   The length of the wire in the magnetic field, L

$$F = BIL \sin \theta$$

Greatest magnetic force occurs when $\theta = 90°$ ($F = BIL$)
Weakest magnetic force occurs when $\theta = 0°$ ($F = 0$)

> **Hint**
>
> **Key formula**
>
> Force on a current-carrying conductor
>
> $F = BIL \sin \theta$
>
> where:
>
> F = magnetic force on the wire (N)
>
> I = current flowing through the wire (A)
>
> B = magnitude of the external magnetic field (T)
>
> θ = angle the wire makes with the external magnetic field (°)
>
> This can also be rearranged to express magnetic field strength as follows:
>
> $$B = \frac{F}{IL \sin \theta}$$

Applying the right-hand rule

The right-hand rule can be used to determine the direction of the magnetic force.

The reason this rule works is that the directions of the magnetic field, the current (velocity of charge) and the electromagnetic force are all perpendicular to each other. It is also notable that they are vector cross products of each other in the x, y, z direction.

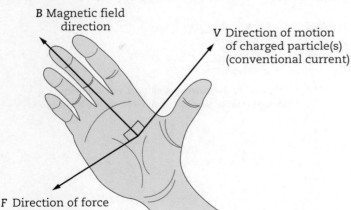

B Magnetic field direction

V Direction of motion of charged particle(s) (conventional current)

F Direction of force

- Right thumb points in the direction of the conventional current or direction of motion of charges (V or I).
- The remaining fingers point in the direction of the external magnetic field (B).
- The palm then points in the direction of the electromagnetic force.

FIGURE 2.14 The right-hand rule identifies the directions of the magnetic field, the moving charge and the electromagnetic force, which are all perpendicular to each other.

Hint

Note: For the flow of negative charges (electrons) the force applied will be in the opposite direction to that of positive charges (conventional current).

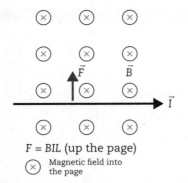

$F = BIL$ (up the page)

$\otimes$ Magnetic field into the page

FIGURE 2.15 Applying the right-hand rule to determine the electromagnetic force.

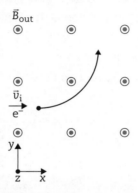

FIGURE 2.16 An electron entering the *B* field in the positive *x* direction.

2.3 Electromagnetic induction

Electromagnetic induction is the production of an electromotice force (emf), or a voltage, in an electrical conductor due to its dynamic interaction with a magnetic field. If this emf acts upon free charges it will induce a current (an **induced current**).

2.3.1 Magnetic flux

Magnetic flux, Φ, is a measurement of the total magnetic field passing through a given area. It is directly proportional to the number of magnetic field lines passing through the defined area (magnetic flux density) as well as the angle at which they pass through. Magnetic flux is measured in webers (Wb).

A change in magnetic flux may be experienced when a magnet is introduced into a coil of wire. This results in the flow of charge (an electric current).

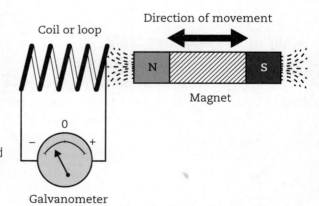

FIGURE 2.17 A current is induced when a magnet is moved relative to a coil of wire connected to a circuit.

> **Hint**
>
> **Key formula**
>
> $$\Phi = B_\perp A = BA \cos \theta$$
>
> where:
>
> Φ = magnetic flux (Wb)
>
> B = magnetic field strength (T)
>
> A = area of the surface (m^2)
>
> θ = angle between the magnetic field lines and a normal to the surface (°)

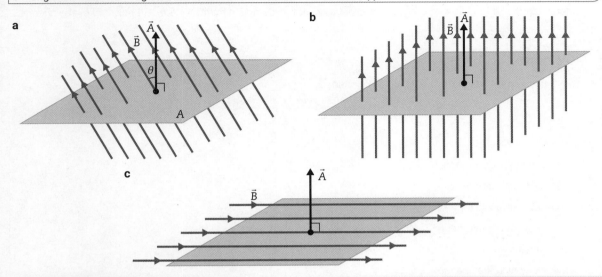

FIGURE 2.18 The flux through the area A depends on the angle, θ, between the field lines and the normal to the plane. In **a** the magnetic flux is reduced as the field lines make an angle with the normal to the surface. In **b** the magnetic flux is maximised as all field lines are parallel to the normal. In **c** the magnetic flux is zero as the field lines do not pass through the area parallel to the normal.

2.3.2 Magnetic flux density

Magnetic flux density, B, is a measure of the strength of a magnetic field. It is the number of magnetic field lines per unit area.

> **Hint**
>
> To recall that magnetic flux density is the number of magnetic field lines in a given area, think about density as packing things into a defined space, hence it is a measure of the number of magnetic field lines per unit area.

2.3.3 Electromotive force

Electromotive force (emf) is a difference in potential that tends to give rise to an electric current. It is measured in volts. Note that it is not actually a force, hence it is not measured in newtons, but rather voltage.

2.3.4 Faraday's law

The magnitude of the induced emf is given by Faraday's law of induction, which states that the **induced emf** in a loop of wire is equal to the negative of the change in magnetic flux (Φ) divided by the change in time (*t*).

Hint

Key formula

Faraday's law of induction:

The induced emf in a loop of wire is equal to the negative of the change in magnetic flux (ΔΦ) divided by the change in time (Δ*t*).

$$\text{emf} = \frac{-\Delta\Phi}{\Delta t} = \frac{-(\Phi_f - \Phi_i)}{\Delta t}$$

Faraday's law and the magnetic flux equation combined:

$$\text{emf} = \frac{-(BA\cos\theta)}{\Delta t} = \frac{-(\Phi_f - \Phi_i)}{\Delta t}$$

where:

emf = electromagnetic force induced (V)

B = magnetic field strength (T)

A = area of the surface (m²)

θ = angle between the magnetic field lines and a normal to the surface (°)

Φ = magnetic flux (Wb)

Δ*t* = time over which the change occurs (s)

The negative sign indicates that the induced current opposes the change in flux (Lenz's law).

To generate a larger emf, a coil containing multiple loops of wire is used. Each loop will have an emf induced between its ends, so connecting *n* loops in series is like connecting *n* batteries in series. Simply add the emf in all loops.

Hint

Key formula

Faraday's law for multiple loops:

$$\text{emf} = \frac{-n(BA\cos\theta)}{\Delta t} = \frac{-n(\Phi_f - \Phi_i)}{\Delta t}$$

where:

emf = electromagnetic force induced (V)

n = number of loops of wire (m⁻¹)

B = magnetic field strength (T)

A = area of the surface (m²)

θ = angle between the magnetic field lines and a normal to the surface (°)

Φ = magnetic flux (Wb)

Δ*t* = time over which the change occurs (s)

A different emf may be generated in a coil of wire by varying any of the variables in the formula

$$\text{emf} = \frac{-n(\Delta BA\cos\theta)}{\Delta t} = \frac{-n(\Phi_f - \Phi_i)}{\Delta t},$$

including the number of loops (*n*), field strength (*B*), area (*A*), angle (*θ*) or time (*t*).

You should practice completing questions by changing any number of these variables.

Worked example

A wire loop of 5 turns and area 0.1 m² experiences a magnetic field strength of 2.0 mT at right angles to it. This field strength reduces from this maximum down to 0.5 mT over a period of 10 seconds. Determine the electromotive force (emf) that is produced over this period.

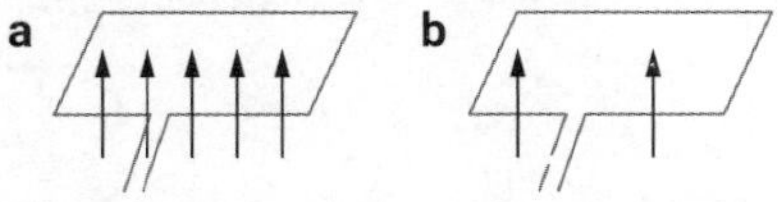

FIGURE 2.19 The flux through the area A of the conducting loop decreases from 2.0 mT in a to 0.5 mT in b over a period of 10 seconds.

Answer

1 State the formula for emf.

$$\text{emf} = \frac{-n(\Delta BA \cos \theta)}{\Delta t} = \frac{-n(\Phi_f - \Phi_i)}{\Delta t}$$

2 Substitute the values into the equation.

$$\text{emf} = \frac{-n(\Delta BA \cos \theta)}{\Delta t} = \frac{-n(\Phi_f - \Phi_i)}{\Delta t}$$

$$\text{emf} = \frac{-5(2.0 \times 10^{-3} \times 0.1 \text{ m}^2 \times \cos 0°) - -5(0.5 \times 10^{-3} \times 0.1 \text{ m}^2 \times \cos 0°)}{10}$$

Note here that the angle, θ, and area, A, remain constant and that the change occurs over a period of 10 seconds.

$$\text{emf} = \frac{-0.001 - -0.00025}{10}$$

$$\text{emf} = -7.5 \times 10^{-5} \text{ V}$$

2.3.5 Lenz's law

Lenz's law states that the direction of an induced electric current always opposes the change in the magnetic field that produces it. This is shown through the use of a negative sign in Faraday's law of induction. As a result, Lenz's law upholds the law of conservation of energy.

Applying the conservation of energy

The potential energy of a changing magnetic field is transformed into electric potential energy. This creates an electric field that can do work by applying a force. This work is done on any free electrons, inducing a current. The induced current takes its energy from the changing magnetic flux, and so reduces the rate at which the flux changes. If this didn't occur, then the flux would increase more, giving a bigger induced current, giving a bigger flux, and so on. This would violate the conservation of energy and thus cannot occur.

a Introducing a
   north pole

When the magnet is moved towards the stationary conducting loop, a current is induced. The magnetic field lines are due to the bar magnet.

This induced current produces its own magnetic field that counteracts the increasing external flux. The right-hand rule shows this.

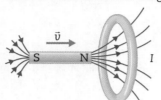

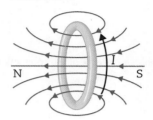

b Removing
   a north pole

When the magnet is moved away from the stationary conducting loop, a current is induced.

This induced current produces a magnetic field that counteracts the decreasing external flux.

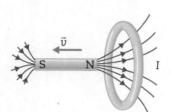

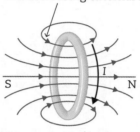

FIGURE 2.20 A moving bar magnet induces a current in a conducting loop. The direction of the current is determined by Lenz's law. **a** Introducing a north pole **b** Removing a north pole.

Hence, Lenz's law is essentially a statement of the conservation of energy.

Example: If a bar magnet was introduced to a coil of wire (a solenoid), the change in flux would produce an induced current. Lenz's law states that this current will counter the change that produced it, thus the magnetic field will oppose change in the magnetic flux.

2.3.6 Electric generators

An **electric generator** transforms kinetic energy into electrical energy. The kinetic energy is in the form of the relative movement of coils of wire and magnets. It does this through the induction of an emf across the coils to generate a current. The coil is attached to an **armature** that rotates in the magnetic field between the poles of the two magnets. As the armature rotates, the flux through the loops of the coil varies, causing an emf across the ends of the coil. Each end of the coil is attached to a conducting slip ring that slides against a brush. The brushes are then connected to the external circuit that uses the emf generated.

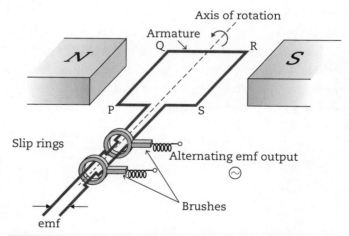

FIGURE 2.21 A schematic diagram of an electric generator.

9780170459174

2.3.7 Electric motors

An electric motor transforms electrical energy into kinetic energy. This is essentially the reverse of an electric generator. An electric motor uses a loop of current-carrying wire in a magnetic field to produce a force. This force allows the loop to rotate and therefore the motor to spin. While both the motor and generator have a similar setup, unlike the generator, the motor contains a split ring commutator. This allows the current to be reversed at the point where the coil is perpendicular to the magnetic field so that the current remains as a **direct current (DC)**.

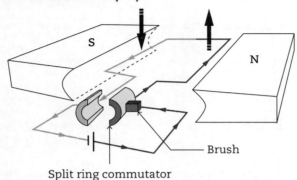

FIGURE 2.22 A schematic diagram of an electric motor. Note the split ring commutator that allows for the current to be direct (one way) in the external circuit.

2.3.8 Transformers

Transformers are devices that use electromagnetic induction to change the voltage of an electric current. A transformer consists of two coils of wire (solenoids) placed near each other and often around a soft **ferromagnetic** iron core.

The primary coil is wound around one side of the iron core, and the secondary coil is wound around the other side. The time-varying or **alternating current (AC)** in the primary coil induces a time varying magnetic field in the secondary coil. Hence, an emf is generated and an AC current is induced in the secondary coil.

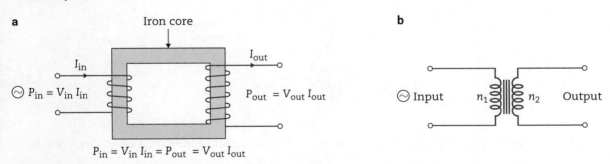

FIGURE 2.23 a A schematic diagram of a transformer and **b** its circuit symbol.

The change in flux is the same for both primary and secondary coils. Hence, transformers are built upon the principles associated with Faraday's law. They have a primary coil and a secondary coil with different numbers of turns. When there's an alternating current in the primary coil, it induces an alternating magnetic field in the surrounding space, which means there's a time-varying magnetic flux through the secondary coil. This generates voltage across the secondary coil and hence induces a current in the secondary coil (electromagnetic induction).

It is important to realise that transformers only work when an alternating current is passing through the primary coil and would not work if direct current was passing through the coil. This is due to the need for a time-changing magnetic field to produce an emf in the secondary coil. This is one of the primary reasons that AC is widely used today – it can easily be transformed.

Transformers can be step down, in which the $V_P > V_S$ and $n_P > n_S$, or step up, in which $V_P < V_S$ and $n_P < n_S$. 'Stepping up' refers to increasing the voltage in the secondary circuit whereas 'stepping down' refers to decreasing the voltage in the secondary circuit. These different types of transformers are termed **step-up** and **step-down transformers**.

> **Hint**
>
> **Key formula**
>
> $$\frac{V_P}{V_S} = \frac{n_P}{n_S}$$
>
> The ratio of the voltages in the two arms of a transformer $\frac{V_P}{V_S}$ is equal to the ratio of the number of coils of each arm $\frac{n_P}{n_S}$.
>
> Assuming that the transformer is 100% efficient, power out = power in or $P_{out} = P_{in}$. We can apply Ohm's law to show that $I_S V_S = I_P V_P$. This can be included in the previous equation to give the full transformer equation.
>
> $$\frac{V_P}{V_S} = \frac{I_S}{I_P} = \frac{n_P}{n_S}$$
>
> The transformer equation: The ratio of the secondary voltage (V_S) to the primary voltage (V_P) is equal to the ratio of the primary current (I_P) to the secondary current (I_S) and is also equal to the ratio of the secondary number of coils (n_S) to the primary number of coils (n_P).
>
> In reality, transformers are not 100% efficient, and a small amount of energy is lost as heat through resistance and **eddy currents**, although this is usually less than 1% of the total energy transformed.

2.4   Electromagnetic radiation

Electromagnetic radiation is defined as radiant energy consisting of synchronised oscillations of electric and magnetic fields, or electromagnetic waves, propagated at the speed of light, c, which is 3.00×10^8 m s^{-1} in a vacuum.

2.4.1 Electromagnetic waves

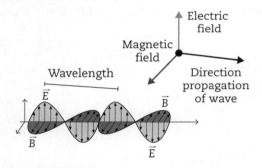

FIGURE 2.24 Electromagnetic waves are transverse waves consisting of coupled electric and magnetic field oscillations.

James Maxwell discovered that electromagnetic waves are produced by an oscillating electric charge resulting in mutually perpendicular electric and magnetic fields. Electromagnetic waves do not require a medium and are therefore able to travel through the vacuum of space. Whenever a charged particle accelerates, it radiates energy in the form of electromagnetic waves.

Electromagnetic spectrum

Maxwell predicted that although visible light was the only kind of electromagnetic wave visible to us, there were many more types of waves across a large range of frequencies and wavelengths.

> **Hint**
> Shorter wavelengths = higher frequency = higher energy waves
> Longer wavelengths = lower frequency = lower energy waves

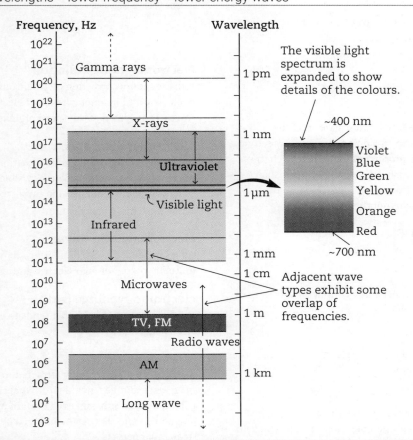

FIGURE 2.25 The electromagnetic spectrum showing all electromagnetic waves that have been produced or detected, ordered according to their wavelengths and frequencies.

Glossary

A+ DIGITAL FLASHCARDS
Revise this topic's key terms and concepts by scanning the QR code or typing the URL into your browser.

https://get.ga/aplus-qce-phys-u34

alternating current (AC)
electrical current that alternates its direction of travel sinusoidally with time

armature
the frame of the rotating part of a motor or generator, holding one or more coils

B field
a magnetic field

Coulomb's law
the second law of electrostatics; states that the force exerted between two point charges is directly proportional to the product of their electric charges, inversely proportional to the square of the distance between them and inversely proportional to the permittivity of the surrounding medium

direct current (DC)
a current that flows in a single direction

eddy current
a circular current induced in a conductor due to a changing magnetic field

electric field
the field due to an electric charge, which applies a force to other electric charges

electric field lines
net lines of force pointing in the direction a positive test charge will move when placed in the electric field due to a charge Q

electric generator
a device used to produce electrical current by electromagnetic induction

electrical permittivity (ε_0)
a physical property of a medium, associated with electricity

electrical potential
potential energy per unit charge in an electric field

electrical potential energy
potential energy stored in an electric field. The change in potential energy of an object is also the work done on that object by the electric field

electromagnet
a magnet with a north and a south pole formed by a current in a solenoid

electromagnetic induction
the production of an electromotive force (emf) or a voltage in an electrical conductor due to its dynamic interaction with a magnetic field

electromagnetic radiation
radiant energy consisting of synchronised oscillations of electric and magnetic fields, or electromagnetic waves, propagated at the speed of light in a vacuum

electromagnetic spectrum
the family of electromagnetic radiations – radio waves, microwaves, infrared radiation, visible light, ultraviolet radiation, X-rays and gamma rays – which travel at $3.0 \times 10^8 \, \mathrm{m \, s^{-1}}$ in a vacuum.

electromagnetic waves
waves produced by an oscillating charge resulting in mutually perpendicular electric and magnetic fields

electromotive force (emf)
a difference in potential that tends to give rise to an electric current

ferromagnetic
materials that are strongly attracted to nearby magnets; examples include iron, nickel and cobalt. Ferromagnetic substances can retain permanent magnetism and this can be induced by other very strong magnets

field
the means by which action-at-a-distance forces are exerted

induced current
a current that is produced due to the presence of an electromotive force

induced emf
an emf created by a changing magnetic field

magnet
a magnetic material that has the majority of its magnetic domains aligned. Magnets have magnetic fields that, in turn, may affect other substances nearby

magnetic domain
a region within a magnetic material (such as iron) where the magnetic properties point in the same direction

magnetic field
the field created by moving charges, including charges in magnetic materials; measured in tesla (T).

9780170459174

magnetic flux
a measurement of the total magnetic field that passes through a given area; has the unit weber (Wb)

magnetic flux density
the strength of a magnetic field per unit area

magnetic force
the force that a magnetic field exerts on a moving charge or current

magnetic permeability (μ_0)
physical property of a medium associated with magnetism

magnetic pole
magnetic north or south pole, a point where magnetic field lines exit or enter

north pole
the pole of a magnet where the external field lines exit

potential difference
the difference in potential between two points in an electric field; the work done per charge

solenoid
a coil of current-carrying wire that creates a large uniform field within the coil

south pole
the pole of a magnet where the external field lines enter

step-down transformer
a transformer with an output potential difference that is lower than the input potential difference, $V_S < V_P$

step-up transformer
a transformer with an output potential difference that is higher than the input potential difference, $V_S > V_P$

uniform electric field
electric field that has the same magnitude and direction at all points

work, W
energy transferred due to the action of a force: $W = Fs$

zero of potential energy
when all charges in the system are infinitely far apart; any other arrangement will have positive or negative potential energy

9780170459174

Revision summary

Use the following summary of syllabus dot points and key knowledge within Unit 3 Topic 2 to ensure that you have thoroughly reviewed the content. Provide a brief definition or comment for each item to demonstrate your understanding or code them using the traffic light system – Green (all good); Amber (needs some review); Red (priority area to review).

The cognitive verbs have been identified in bold.

Electrostatics	
• **define** Coulomb's law and recognise that it describes the force exerted by electrostatically charged objects on other electrostatically charged objects	
• **solve** problems involving Coulomb's law	
• **define** the terms *electric fields, electric field strength* and *electrical potential energy*	
• **solve** problems involving electric field strength	
• **solve** problems involving the work done when an electric charge is moved in an electric field	
Magnetic fields	
• **define** the term *magnetic field*	
• **recall** how to represent magnetic field lines, including sketching magnetic field lines due to a moving electric charge, electric currents and magnets	
• **recall** that a moving electric charge generates a magnetic field	
• **determine** the magnitude and direction of a magnetic field around electric current-carrying wires and inside solenoids	
• **solve** problems involving the magnitude and direction of magnetic fields around a straight electric current-carrying wire and inside a solenoid	
• **recall** that electric current-carrying conductors and moving electric charges experience a force when placed in a magnetic field	

»

»

• **solve** problems involving the magnetic force on an electric current-carrying wire and moving charge in a magnetic field	
• Mandatory practicals: – **Conduct** an experiment to investigate the force acting on a conductor in a magnetic field – **Conduct** an experiment to investigate the strength of a magnet at various distances	

Electromagnetic induction

• **define** the terms magnetic *flux*, *magnetic flux density*, *electromagnetic induction*, *electromotive force (EMF)*, *Faraday's law* and *Lenz's law*	
• **solve** problems involving the magnetic flux in an electric current-carrying loop	
• **describe** the process of inducing an emf across a moving conductor in a magnetic field	
• **solve** problems involving Faraday's law and Lenz's law	
• **explain** how Lenz's law is consistent with the principle of conservation of energy	
• **explain** how transformers work in terms of Faraday's law and electromagnetic induction	

Electromagnetic radiation

• **define** and **explain** electromagnetic radiation in terms of electric fields and magnetic fields	

© State of Queensland (QCAA) Physics General Senior Syllabus 2019 CC BY 4.0 https://creativecommons.org/licenses/by/4.0/

Exam practice

Multiple-choice questions

Each multiple-choice question is worth 1 mark.

Solutions start on page 125.

Question 1

The diagram shows two conducting coils linked by a changing magnetic flux. Coil A is connected to a circuit that includes a battery, a switch, and a resistor. Coil B lies in the same plane but is not connected to coil A.

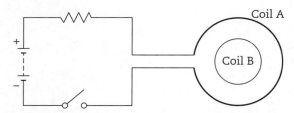

The direction of the induced current in coil B at the moment when the switch is closed in coil A's circuit can be described as:

A anticlockwise.

B anticlockwise to clockwise.

C clockwise.

D there will be no induced current.

Question 2

Two small spheres have equal charges q and are separated by a distance r. The force exerted on each sphere by the other has magnitude F.

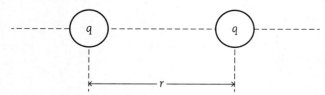

Should the distance between them, r, be tripled, then the magnitude of the force on each sphere is reduced to:

A $3F$

B $\dfrac{F}{3}$

C $\dfrac{F}{9}$

D $9F$

Question 3

Identify which of the following best describes the term 'electromagnetic radiation'.

A a magnetic field created by moving charges

B energy that travels in waves represented by the colours of the visible spectrum

C the waves or photons of an electromagnetic field, radiating through space and carrying radiant electromagnetic energy

D waves of light across the spectra of visible light, ultraviolet light, X-rays and gamma radiation

Question 4

A wire with mass m kg and length L m has a conventional current I amperes flowing to the left.

The conducting wire is suspended against the gravitational field, that is, the magnetic force cancels out the gravitational force on the wire. Infer the direction of the external magnetic field.

A down the page

C out of the page

B into the page

D up the page

Question 5

> **Hint**
>
> When using the formula for the number of turns in a solenoid be sure to find the value n as the number of turns per metre length, $n = \dfrac{N}{L}$, where n is the number of turns per metre, N is the number of turns and L is the length in metres.

The number of turns of wire in a solenoid is unknown but may be determined given the other variables. Given a magnetic field strength inside the solenoid, B, of 4.0×10^{-5} T, a length of 30 cm and a current of 1.8 A running through the solenoid, then the number of turns of wire may be determined to be:

A 1 turn.

C 18 turns.

B 5 turns.

D 30 turns.

Question 6

Identify which of the following statements is the best description of the process of electromagnetic induction.

A a difference in electrical potential that gives rise to an electric current

B energy can neither be created nor destroyed, but only changed from one form to another or transferred from one object to another

C the resulting increase in electric current produced by a constant magnetic field

D the production of an electromotive force (emf) or voltage across an electrical conductor due to its interaction with a changing magnetic field

Question 7

The wavelength of a microwave of frequency 4.8×10^{10} Hz is closest to:

A 1.60 mm

C 0.625 m

B 6.25 mm

D 1.6 m

Question 8

Indicate the direction of the induced current in the conducting loop below.

A clockwise, as viewed from the right of the page

B clockwise, as viewed from the left of the page

C counter-clockwise, as viewed from the left of the page

D counter-clockwise, as viewed from the right of the page

Question 9

A current-carrying wire in an external magnetic field experiences a force.

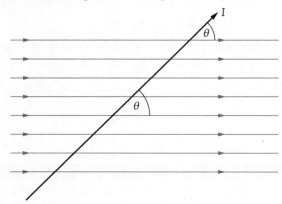

The direction of this force relative to a conventional current would be described as:

A at an angle θ to the direction of the field.

B into the page.

C out of the page.

D perpendicular to the angle θ.

Question 10

Consider a system of two charges, Q_1 and q_2. If $Q_1 = +20\,\mu C$ and $q_2 = -10\,\mu C$, and the separation distance is 20 cm, then the calculated force between the two charges would be equal to:

A $-50\,N$

B $-45\,N$

C $-4500\,N$

D $+5000\,N$

Short response questions

Question 11 (4 marks)

a Explain how transformers work in terms of Faraday's law and electromagnetic induction.   3 marks

> **Hint**
>
> When answering a question asking you to *explain*, remember to follow the QCAA definition.
>
> **explain**: make an idea or situation plain or clear by describing it in more detail or revealing relevant facts; give an account; provide additional information

b A transformer is labelled as 240 V AC input and 60 V AC output. It has 80 turns in its primary coil. Calculate the number of windings in the secondary coil of the transformer.   1 mark

Question 12 (4 marks)

a Determine the direction of the current in the loop shown below (in terms of X and Y) when the field direction in the loop is reversed from up the page to down the page.   2 marks

Hint

When answering a question asking you to *determine*, remember to follow the QCAA definition.

determine: establish, conclude or ascertain after consideration, observation, investigation or calculation; decide or come to a resolution.

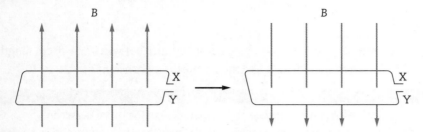

b If the magnetic field strength reduces from $B = 0.10\,\text{T}$ to $B = -0.10\,\text{T}$ over a period of 2 seconds, determine the magnetic flux that the $8\,\text{cm} \times 24\,\text{cm}$ loop experiences.   2 marks

Question 13 (4 marks)

A negatively charged particle is fired into a magnetic field and undergoes motion as shown below.

a Describe the direction of the magnetic field through which the negatively charged particle is moving.   2 marks

Hint

When answering a question asking you to *describe*, remember to follow the QCAA definition.

describe: give an account (written or spoken) of a situation, event, pattern or process, or of the characteristics or features of something

b Determine what would occur to the radius of the arc travelled when:

   i the velocity of the charged particle increased by a factor of 3.   1 mark

   ii the magnitude of the charge on the particle doubled.   1 mark

Question 14 (4 marks)

In an electric field about a point charge, the electric field strength is $18\,\text{N}\,\text{C}^{-1}$ at point R, 1 metre from the fixed charge, $+q$.

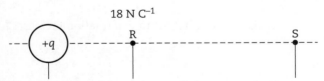

a Determine the electric field strength at point S, a distance 4 metres from the fixed charge.   2 marks

b Calculate $+q$ and hence determine the distance from the fixed charge where the electric field strength would be $10\,\text{N}\,\text{C}^{-1}$.   2 marks

Question 15 (3 marks)

A proton of mass $1.67 \times 10^{-27}\,\text{kg}$ and charge $+1.6 \times 10^{-19}\,\text{C}$ enters at right angles to a magnetic field of strength $B = 3.0 \times 10^{-2}\,\text{T}$ into the page and with a velocity of $2.5 \times 10^{5}\,\text{m}\,\text{s}^{-1}$.

a Calculate the magnitude of the force on the proton.   1 mark

b Sketch a diagram to show its motion in an external magnetic field. 1 mark

> **Hint**
>
> When answering a question asking you to *sketch*, remember to follow the QCAA definition.
>
> **sketch**: execute a drawing or painting in simple form, giving essential features but not necessarily with detail or accuracy; in mathematics, represent by means of a diagram or graph; the sketch should give a general idea of the required shape or relationship and should include features

c Calculate the centripetal acceleration of the proton. 1 mark

Question 16 (2 marks)

Calculate the field strength inside a solenoid made up of 600 turns of wire on a 40 cm length of plastic tube if the wire carries a current of 500 mA. Use the permeability constant of $1.26 \times 10^{-6}\,\text{T A}^{-1}\text{m}$.

> **Hint**
>
> When answering a question asking you to *calculate*, remember to follow the QCAA definition.
>
> **calculate**: determine or find (e.g. a number, answer) by using mathematical processes; obtain a numerical answer showing the relevant stages in the working; ascertain/determine from given facts, figures or information

> **Hint**
>
> When performing calculations for solenoids, the number of turns, n, is a number per metre length, hence the actual number of turns, N, needs to be divided by the length of the solenoid, L, in metres.
>
> $$n = \frac{N}{L} \text{ turns per metre}$$

Question 17 (2 marks)

The graph below shows the output of an electrical generator that is rotating at 50 Hz in a magnetic field of 0.30 T.

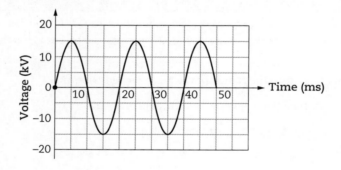

Describe the changes to the output waveform when:

a the magnetic field doubles to 0.60 T 1 mark

b the magnitude of the field remains the same but the rate of rotation increases to 100 Hz. 1 mark

Question 18 (4 marks)

A wire of length 40 cm is placed entirely in a magnetic field of strength 0.6 T and carries a current of 1.5 A. The conducting wire experiences a force of 0.18 N.

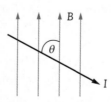

a Determine the angle θ it must make with the field. 3 marks

b State the direction of the force, justifying your reasoning. 1 mark

Question 19 (3 marks)

Calculate the net force acting on the charged object B due to charge A and charge C as in the arrangement below.

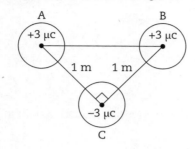

Question 20 (5 marks)

A coil of wire is suspended from a spring balance between two poles of a pair of magnets. The coil is 15 cm high and 4 cm wide and has 25 turns of wire.

In the experiment, the spring balance readings were recorded for a range of different currents. The apparatus and results are shown below.

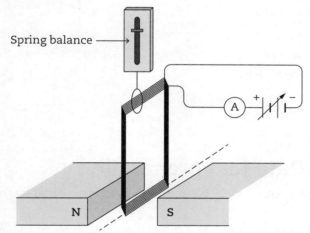

Current (A)	Mass (g)	Force (N)
0.5	128	1.3
1.0	149	1.5
1.5	173	1.7
2.0	194	1.9
2.5	212	2.1
3.0	235	2.3
3.5	255	2.5

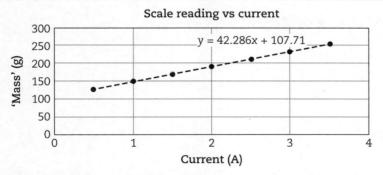

a Determine the mass and weight of the coil using the table of data and graph provided.   2 marks

b Calculate the magnetic field strength with reference to the gradient of the graph.   1 mark

c The current is adjusted such that the balance now reads zero. Determine the current that now flows in the coil and state the direction that it must flow.   2 marks

Chapter 3
Unit 3 Data test

The Data test is the first of the summative internal assessments and is specifically associated with Unit 3 content.

It addresses Assessment Objectives 2, 3 and 4, requiring you to apply understanding, and analyse and interpret evidence. It is completed individually, under supervised conditions, with 60 minutes of working time and 10 minutes of perusal time.

	The response must be:	Cognitive verbs:
Apply understanding	An unknown scientific quantity or feature.	Calculate (show your working), identify, recognise, use evidence.
Analyse evidence	A trend, pattern, relationship, limitation or uncertainty in the data sets.	Categorise, classify, contrast, distinguish, organise, sequence.
Interpret evidence	A conclusion based on the data sets (not your general knowledge).	Compare, deduce, extrapolate, infer, justify, predict.

TABLE 3.1 Summary of the types of responses possible in IA1.

3.1 Guidance for Data test preparation

There are some data test specific preparations that you can employ to ensure that you optimise your performance.

The Data test is a specific type of assessment item with a characteristic format. Focus content and conceptual areas and typical question types are noted below. This assessment task is primarily focussed on responding to items using qualitative data and/or quantitative data derived from the mandatory or suggested practicals, activities or case studies from the unit being studied. You are expected to respond to the following types of questions:

*   short response items requiring single-word, sentence or short-paragraph responses
*   calculating using algorithms
*   interpreting datasets.

3.1.1 Unit 3 content and concepts

*   Vectors
*   Projectile motion
*   Inclined planes
*   Circular motion
*   Gravitational force and fields
*   Orbits
*   Electrostatics
*   Magnetic fields
*   Electromagnetic induction
*   Electromagnetic radiation

3.1.2 Typical question types

- Calculate an average value from a table or set of values
- Read a value from a graph
- Determine an absolute uncertainty
- Perform a general calculation using a formula provided
- Identify trends or relationships from a table or graph (e.g. linear, quadratic, inverse square)
- Predict the magnitude of a value with a given value and the relationship
- Calculate the percentage error (between a theoretical and an experimental value)
- Sketch a graph of an equation
- Use the data in a table or read from a graph to determine any trends (and justify them with specific values)
- Draw a conclusion that quantifies the magnitude of a variable, including the absolute or percentage uncertainty

3.1.3 Data test tips

- If provided with a table of data and asked to calculate a quantity that could come from all of the data points, e.g. specific heat capacity, you should always graph the data to gain the equation and use the gradient so that all points are considered, rather than choosing a single point from the table.
- When identifying relationships, be as specific as possible. The data will allow you to move beyond just using the words increase, decrease. There could be a variety of relationships – linear, quadratic, inverse, etc. If you know a relationship to be directly proportional, then state this.
- The data you reference to support statements needs to be specific. If you are saying it is a linear relationship you must outline that the points are increasing at a constant rate to support this, rather than just provide two points that demonstrate an increase.
- If you require a previous answer to complete a question that is asking you to demonstrate a skill (that you haven't been able to complete), use a reasonable value as a place holder so that you can still demonstrate the skill and receive some credit. For example, you can use any value (or a value from the *Physics formula and data book*) to apply the rule for absolute uncertainty, even if you couldn't answer a preceding question.
- Pay attention to the units in the question and the units provided in the answer box. They must be compatible, and you must respond with the units provided.
- Address the question asked and the cognitive verb in the question stem.
- If a question asks you to use a graph in the response then you must use the trendline from the graph to gain full marks, even though there may be other mathematically valid ways of responding (such as simply inputting a value into a formula).
- If it asks you to use the word **determine** you are required to find the value yourself, rather than referring to a table of known constants or values.
- It is important to match the type of response to the verb in the question.

Assessment objective	Verb	Definition (QCAA)
Apply	Calculate	Determine or find (e.g. a number, answer) by using mathematical processes; obtain a numerical answer showing the relevant stages of working; ascertain/determine from given facts, figures or information.
	Identify	Distinguish; locate, recognise and name; establish or indicate who or what someone or something is; provide an answer from a number of possibilities; recognise and state a distinguishing fact or figure.
	Recognise	Identify or recall particular features of information from knowledge; identify that an item, characteristic or quality exists; perceive as existing or true; be aware of or acknowledge.
	Use evidence	Operate or put into effect; apply knowledge or rules to put theory into practice.
Analyse	Categorise	Place in or assign to a particular class or group; arrange or order by classes or categories; classify, sort out, sort, separate.
	Classify	Arrange, distribute or order in classes or categories according to shared qualities or characteristics.
	Contrast	Display recognition of differences by deliberate juxtaposition of contrary elements; show how things are different or opposite; give an account of the differences between two or more items or situations, referring to both or all of them throughout.
	Distinguish	Recognise as distinct or different; note points of difference between; discriminate; discern; make clear difference/s between two or more concepts or items.
	Organise	Arrange, order; form as or into a whole consisting of interdependent or coordinated parts, especially for harmonious or united action.
	Sequence	Place in a continuous or connected series; arrange in a particular order.
Interpret	Compare	Display recognition of similarities and differences and recognise the significance of these similarities and differences.
	Deduce	Reach a conclusion that is necessarily true, provided a given set of assumptions is true; arrive at, reach or draw a logical conclusion from reasoning and the information given.
	Extrapolate	Infer or estimate by extending or projecting known information; conjecture; infer from what is known; extend the application of something (e.g. a method or conclusion) to an unknown situation by assuming that existing trends will continue or similar methods will be applicable.
	Infer	Derive or conclude something from evidence and reasoning, rather than from explicit statements; listen or read beyond what has been literally expressed; imply or hint at.
	Justify	Give reasons or evidence to support an answer, response or conclusion; show or prove how an argument, statement or conclusion is right or reasonable.
	Predict	Give an expected result of an upcoming action or event; suggest what may happen based on available information.

© State of Queensland (QCAA) Physics General Senior Syllabus 2019 CC BY 4.0https://creativecommons.org/licenses/by/4.0/
TABLE 3.2 Definitions of cognitive verbs associated with the Data test assessment objectives.

9780170459174

3.1.4 Final Data test preparations

It is recommended that you:

- Refer to your copy of the *Physics formula and data book*.

- Refer to the QCAA Sample assessment instruments and Sample marking schemes.

- Review the 'Data test tips' above.

- Look back to any previous formative assessment Item 1 Data tests and the specific feedback provided by your teacher.

- Revise both the content and the concepts from completed Practical Experiments and any analyses of graphs over the course studied.

- Review how to analyse the relationship between any two variables from a linear graph. Use the relevant formula and the variables on each axis to identify what the gradient represents.

- Remind yourself of the typical question types that you can expect to see in a Data test. This includes reviewing and understanding the data and analyses from each of the mandatory practical experiments.

3.1.5 Data test conditions

- Time: 60 minutes plus 10 minutes perusal.

- Length: up to 500 words in total, consisting of

    - short responses, i.e. single words, sentences or short paragraphs (fewer than 50 words)

    - paragraphs, 50–250 words per item

    - other types of item responses (e.g. interpreting and calculating) that should allow students to complete the response in the set time.

- Other:

    - QCAA-approved graphics calculator permitted

    - *Physics formula and data book* permitted

    - unseen stimulus, typically Data sets.

Data set 1

Solutions start on page 129.

A student set up a gravitational force experiment based on the Cavendish experiment to measure the force exerted by a 50 kg mass on a 1 kg mass when separated by distances (d) between 0.010 m ≤ d ≤ 0.060 m. The raw data from this experiment is presented in the table below.

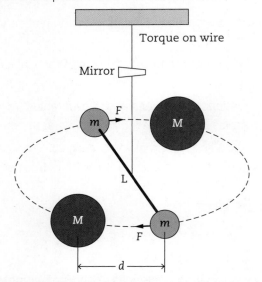

FIGURE 3.1 Apparatus for the Cavendish gravitational force experiment.

Distance d (m) ±0.0005 m	Force F (μN) ± 0.01 μN			Average force F (μN)
	Trial 1	Trial 2	Trial 3	
0.010	33.34	33.76	33.10	33.40
0.020	8.35	8.52	8.31	
0.030	3.70	3.57	3.85	3.71
0.040	2.10	2.22	2.08	2.13
0.050	1.35	1.31	1.50	1.39

TABLE 3.3 Results from gravitational force experiment.

Question 1: Apply understanding   (1 mark)

Calculate the average force (μN) for the distance d = 0.020 m.

Question 2: Apply understanding   (2 marks)

Determine the absolute uncertainty of the mean for the force, F, between the masses when separated

by a distance d = 0.020 m. Use the formula $\pm \dfrac{x_{max} - x_{min}}{2}$.

Question 3: Analyse evidence   (2 marks)

Identify the relationship between the distance, d, and the force, F, exerted. Use evidence from Table 3.3 to support your answer.

Question 4: Interpret evidence   (3 marks)

Predict the magnitude of the force between the two masses when they are 0.080 m apart. Show your working.

Data set 2

Solutions start on page 130.

An experiment was conducted to explore the relationship between the initial velocity ($1.0\,\text{m/s} \le u \le 10.0\,\text{m/s}$) of a ball and its maximum height when launched from ground level with a constant projection angle of 40°.

The experimental data collected is presented in the graph below.

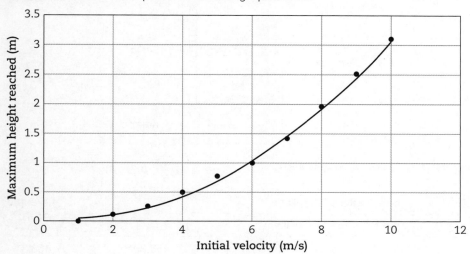

FIGURE 3.2 Maximum height of a ball projected with various initial velocities.

Question 1: Apply understanding   (2 marks)

Calculate the horizontal component of the velocity of the ball when it has an initial velocity of $10.0\,\text{m s}^{-1}$.

Give the answer correct to 2 decimal places.

Question 2: Apply understanding   (1 mark)

Identify the maximum height reached when the ball was projected with an initial velocity of $4.0\,\text{m s}^{-1}$.

Give the answer correct to 1 decimal place.

Question 3: Analyse evidence   (2 marks)

Identify the relationship between the initial velocity, u, and the maximum height of the ball. Use specific values to justify your response.

Question 4: Interpret evidence   (2 marks)

Infer the maximum height that would be expected if the initial velocity was $12.0\,\text{m s}^{-1}$. Give a mathematical reason for your response.

Data set 3

Solutions start on page 131.

An experiment to identify the relationship between the force exerted by a charged particle on another identical charged particle and the distance of separation between their centres (*d*) for $0.025 \, \text{m} \leq r \leq 0.125 \, \text{m}$.

The Coulomb's law experimental data is presented in the table below.

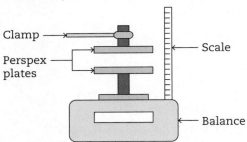

FIGURE 3.3 Apparatus for Coulomb's law experiment.

Distance *d* (m) ±0.0005 m	Force *F* (μN) ± 0.01μN			Average force *F* (μN)
	Trial 1	Trial 2	Trial 3	
0.025	453.12	453.00	453.80	453.31
0.050	115.32	115.39	115.00	115.24
0.075	63.12	62.98	63.20	63.10
0.100	28.60	28.11	27.99	
0.125	20.88	21.24	21.24	22.12

TABLE 3.4 Results from electrostatic force experiment.

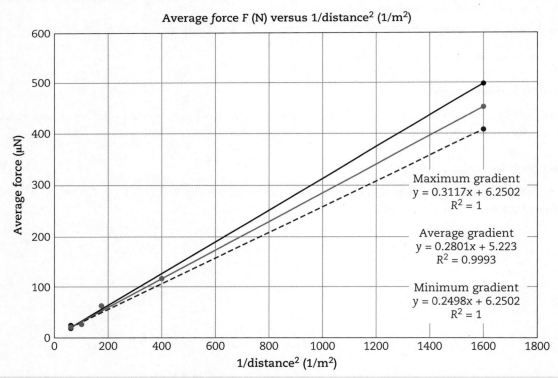

FIGURE 3.4 Graph of average force (N) versus $\dfrac{1}{\text{distance}^2} \left(\dfrac{1}{\text{m}^2} \right)$ for the Coulomb's law experiment.

Question 1: Apply understanding (2 marks)

Calculate the uncertainty in the gradient and the percentage uncertainty in the gradient of the average force (N) versus $\dfrac{1}{\text{distance}^2}\left(\dfrac{1}{\text{m}^2}\right)$ graph.
Give answers correct to 3 significant figures.

Question 2: Analyse evidence (2 marks)

Identify the relationship between the distance, d, between two charges and the force, F, exerted by one charge on the other. Use evidence from the table in the dataset to support your answer.

Question 3: Interpret evidence (3 marks)

Predict the magnitude of the force between the two charges when they are 0.250 m apart.

Show your working and give your answer correct to 2 decimal places.

UNIT 4
REVOLUTIONS IN MODERN PHYSICS

9780170459174

Chapter 4
Topic 1: Special relativity

Topic summary

Newton's early work on motion had been found to be valid in all physical circumstances until the twentieth century where small, but significant, discrepancies were found between predictions based on Newtonian physics and experimental results. Refinements to the theory of electromagnetism were a trigger for this change, leading to the development of the special theory of relativity. 'Special relativity' contrasts Newtonian physics with Albert Einstein's new way to look at the nature of time, space, matter and energy. Concepts of inertial frames of reference, simultaneity, time dilation, length contraction and relativistic momentum are explored and the postulates of special relativity and paradoxical scenarios are used to explain the phenomena of special relativity.

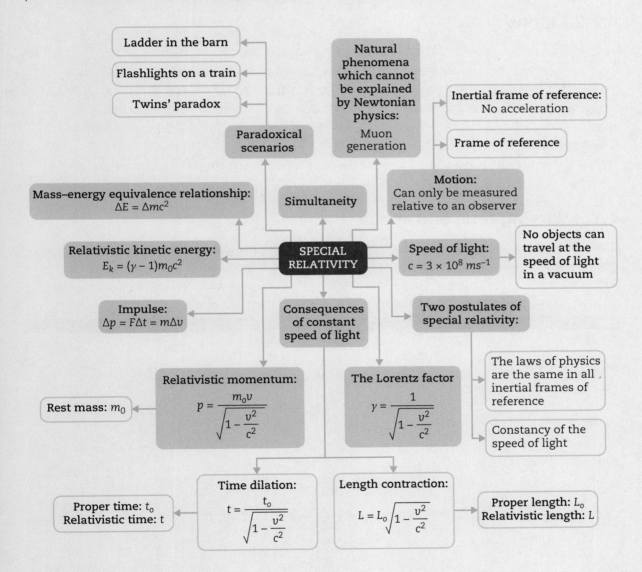

4.1 Special relativity

4.1.1 Phenomena unable to be explained by Newtonian physics

Prior to Einstein and **special relativity**, the universe was understood in terms of Newton's three laws of motion:

1 Objects in motion (or at rest) remain at constant motion unless acted upon by an external force.

2 Force is equal to the change in momentum per change of time. For a constant mass, force equals mass times acceleration.

3 For every action, there is an equal and opposite reaction.

However, some observed phenomena did not obey **classical mechanics** and could only be explained using **quantum physics** and special relativity, not Newtonian physics. **Muons** are one such example of a natural phenomena that cannot be explained by Newtonian physics.

4.1.2 Muons

Muons, while similar to electrons, are unstable so they have a short mean life-time (only 2.2 microseconds). Muons are often observed travelling toward Earth at speeds close to the speed of light. Due to their short mean life-time, muons should typically decay before reaching the surface of Earth, according to Newtonian physics. In contrast, experiments show that a large majority of the muons detected in the upper atmosphere are also detected close to the Earth's surface.

These experiments suggest the muons undergo **relativistic effects**, thereby supporting special relativity.

4.1.3 Varying time and space

Time and space act differently for one object compared to another depending on their relative speed, when the objects are travelling near the speed of light. Objects which experience these relativistic effects do not obey the time and space constraints of Newtonian physics. In fact, relative to an observer, length can contract (get smaller), and time can dilate (stretch).

Hint	
Key formula	
where t = relativistic time t_0 = proper time	$t = \dfrac{t_0}{\sqrt{1 - \dfrac{v^2}{c^2}}}$
where L = relativistic length L_0 = proper length	$L = L_0 \sqrt{1 - \dfrac{v^2}{c^2}}$
where γ = the Lorentz factor	$\gamma = \dfrac{1}{\sqrt{1 - \dfrac{v^2}{c^2}}}$
where p = relativistic momentum m_0 = rest mass	$p = \dfrac{m_0 v}{\sqrt{1 - \dfrac{v^2}{c^2}}}$
	$E = mc^2$

4.1.4 The Lorentz factor

The **Lorentz factor** is a factor by which both time dilation and length contraction are affected when velocity approaches the speed of light, c.

$$\gamma = \frac{1}{\sqrt{1 - \dfrac{v^2}{c^2}}}$$

where γ = the Lorentz factor

This is a common element for determining relativistic time, length and momentum.

4.1.5 Frames of reference

A **frame of reference** is a framework in which the motion of an object is described according to a coordinate system. Frames of reference are observational and can be inertial or non-inertial.

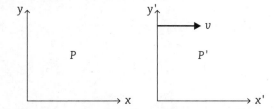

FIGURE 4.1 Two inertial frames that are moving relative to each other. *P* is regarded as the stationary frame. Frame *P'* has a velocity *v* in the *x*-direction (relative to *P*).

An **inertial frame of reference** is one in which Newton's first law applies and there is no acceleration. It can be stationary, or moving at a constant velocity.

Newton's laws of motion are the same in any inertial frame of reference, and the laws of conservation of energy and momentum also apply. All inertial frames of reference are equally valid.

In his book *Dialogues on the Two Chief Systems of the World*, Galileo described a thought experiment in which a sailor drops an object from the mast of a sailing ship moving at a steady velocity. He asked the question: 'Where would the object land relative to the deck of the ship?' In his frame of reference, the sailor would see the object fall straight down parallel to the mast (Figure 4.2a); however, a nearby observer on land (a different frame of reference) would see the object follow a parabolic path (Figure 4.2b).

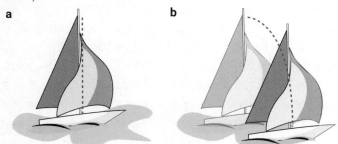

FIGURE 4.2 Path of a falling object in the reference frame of (a) a sailor on the ship and (b) an observer on land.

4.1.6 Postulates of special relativity

Einstein's theory of special relativity was based on the following two propositions:

First postulate of special relativity: the laws of physics are the same in all inertial frames of reference – the principle of special relativity.

Second postulate of special relativity: the speed of light has the same value, c, in all inertial frames. It does not depend on the speed of either the source or the observer.

4.1.7 Simultaneity

Remember: motion can only be measured relative to an observer.

When two events occur simultaneously, they are said to happen at the same time. However, something considered a simultaneous event in one inertial reference frame may not be considered simultaneous in another.

Simultaneity occurs when one reference frame is moving relative to another reference frame, and events that occur simultaneously in one reference frame are not considered to be simultaneous in the other.

Now let's get an idea of how the concept of simultaneity works, or how something that might appear simultaneous to one observer may not appear simultaneous to another. Consider the case where two identical street lamps are turned on at the same time. Does this event appear simultaneous?

Well, to one observer located equally distant from both lamps, the answer is yes, the events are simultaneous.

But to another observer, located closer to one lamp than the other, the answer is no, the events are not simultaneous.

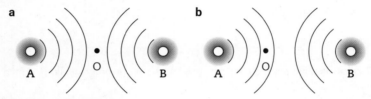

FIGURE 4.3 Reference frames for street lamps turning on

In Figure 4.3a from the position of the stationary observer, O, the turning on of the street lamps appears to be simultaneous. In Figure 4.3b, from the position of the observer, who is moving west, the turning on of the street lamps does not appear to be simultaneous (they see lamp A turn on before lamp B) due to their motion west.

This leads to the statement that time is not absolute, i.e. that two events that are simultaneous to one observer are not necessarily simultaneous to another observer in a different reference frame.

4.1.8 Time dilation

Time dilation dictates that a longer time may be measured by an observer outside the reference frame in which an object is observed to be moving; for example, moving clocks appear to run more slowly. QCAA defines time dilation as the difference of elapsed time between two events as measured by observers moving relative to each other.

Proper time, t_0, is the time measured between two events occurring at the same place in an inertial reference frame, as measured by an observer in that inertial reference frame.

The **relativistic time**, t, is the time measured in the frame of reference in which the object is considered to be in motion (e.g. a stationary observer on Earth watching a moving spaceship).

Time dilation can be calculated using the formula:

$$t = \frac{t_o}{\sqrt{1 - \dfrac{v^2}{c^2}}}$$

where t = time measured from an observer outside the frame of reference (the object is observed to be travelling close to the speed of light)

t_0 = time measured from an observer inside the frame of reference (the object is observed to be stationary)

v = the speed of the object

c = the speed of light

9780170459174

4.1.9 Length contraction

Length contraction means that length measurements are shorter in a reference frame that is moving relative to an inertial frame; that is when the length of an object travelling close to the speed of light is measured, it will appear 'contracted'.

Proper length, L_0, is the length measured in the frame of reference in which the object is at rest. It is the true length of the object without any relativistic effects.

The **relativistic length**, L, is the length measured in the frame of reference in which the object is considered to be in motion.

Length contraction can be calculated using the following formula:

$$L = L_0 \sqrt{1 - \frac{v^2}{c^2}}$$

where L = relativistic length
L_0 = proper length
v = the speed of the object
c = the speed of light

Worked example

At what velocity would a 1 m-long ruler have to move relative to an observer to be measured as being only 0.5 m long by the observer?

Answer

1 State the length contraction formula.

$$L = L_0 \sqrt{1 - \frac{v^2}{c^2}}$$

2 Identify the variables.

L = 0.5 m (the relativistic length)

L_0 = 1.0 m (the proper length)

3 Substitute the values into the formula.

$$L = L_0 \times \sqrt{1 - \frac{v^2}{c^2}}$$

$$\frac{L}{L_0} = \sqrt{1 - \frac{v^2}{c^2}}$$

$$\frac{0.5}{1.0} = \sqrt{1 - \frac{v^2}{c^2}}$$

$$0.25 = 1 - \frac{v^2}{c^2} \text{ (square both sides)}$$

$$\frac{v^2}{c^2} = 0.75$$

Therefore $v^2 = 0.75 \times c^2$

And $v = \sqrt{0.75} \times c$

So $v = 0.87\,c$

4.1.10 Relativistic momentum

Relativistic momentum is the momentum of a particle due to its high relative speeds.

It is measured using the formula:

$$p_v = \frac{m_o v}{\sqrt{1 - \dfrac{v^2}{c^2}}}$$

where p_v = relativistic corrected momentum
m_0 = rest mass
v = velocity of the object relative to the observer
c = the speed of light

4.1.11 Mass–energy equivalence

The famous equation $\Delta E = \Delta mc^2$ is termed 'Einstein's mass–energy equivalence relationship'. According to this equation, 1 kg mass of any matter is equivalent to 9×10^{16} J of energy. This is a huge amount of energy, as the value of c is very high. This equation is taken advantage of by nuclear fission power reactors to provide energy to many millions of people worldwide.

This mass–energy equivalence relationship is the connecting link between energy and matter – that is, mass can appear as energy, and energy can appear as mass.

Worked example

Calculate the rest energy, in joules, of an electron of mass 9.11×10^{-31} kg using the mass–energy equivalence formula.

Answer

1 State the mass–energy equivalence formula.

$$\Delta E = \Delta mc^2$$

2 Substitute in the known values.

$$E = 9.11 \times 10^{-31} \times (3.0 \times 10^8)^2$$

$$E = 8.20 \times 10^{-14} \text{ J}$$

The mass of an electron of 9.11×10^{-31} kg is equivalent to 8.20×10^{-14} joules of energy.

4.1.12 Evidence of special relativity

One example of evidence supporting special relativity already discussed is the proportion of muons reaching the surface of Earth. Relativistic effects such as length contraction and time dilation allow muons with their dilated mean life-times to reach the Earth's surface, supporting special relativity.

Can we reach the speed of light?

As an object moves faster, its relativistic momentum increases (and its mass may appear to increase) while its length contracts. This means that at the speed of light, an object's momentum is infinite. As $E = mc^2$, this would require an infinite amount of energy. Therefore, it is impossible for an object to reach the speed of light in a vacuum.

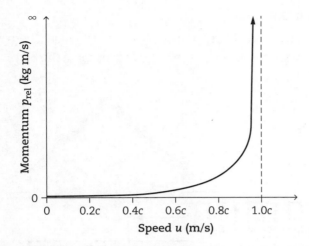

FIGURE 4.4 As the velocity of an object increases, its relativistic momentum approaches infinity, hence c, the speed of light, is considered a cosmic speed limit.

4.1.13 Paradoxical situations

A paradox is a situation where an outcome leads to a conclusion that seems logically unacceptable or self-contradictory. In this course, there are three seemingly paradoxical situations related to special relativity: the twins' paradox, flashlights on a train, and the ladder in the barn paradox.

The twin's paradox

The twin's paradox occurs when one identical twin, Twin B, takes off in a spaceship experiencing time dilation. When Twin B returns to Twin A, Twin A is 10 years older but views Twin B as one year older – time dilation meant that time 'slowed down' for Twin B.

However, each twin sees the other twin moving relative to their own reference frame, so they could both make a similar claim. Hence, paradoxically, each twin should have found the other to have aged less as inertial reference frames are all considered equally valid.

Resolution

The twin's paradox can be solved by taking into account the journey of each of the twins, and what defines an inertial reference frame. An inertial reference frame does not accelerate, and Twin B taking off to space and returning undergoes both acceleration in leaving and deacceleration when returning. Hence, Twin A is considered to measure the 'proper time'.

Flashlight on a train paradox

Consider a moving train in which there are two people, one at each end of a carriage, and there is also an observer on the ground. The person at the back of the carriage shines a flashlight toward the other end of the carriage.

The person at the front of the carriage would observe the light travelling at $v = c$; however, the person on the ground would observe the light travelling at $v + c$. This is paradoxical, as nothing can travel faster than c.

Resolution

This problem can be solved by considering the laws of special relativity.

Einstein's second postulate states that the speed of light has the same value in all inertial frames, regardless of the speed of the source or the observer. Hence, the speed of light is invariant.

As the observer on the platform observes the train moving, and the speed of light is invariant, then the resolution is based upon relativistic contracted lengths or dilated times being observed.

The ladder in the barn paradox

This paradoxical scenario involves a ladder that doesn't quite fit inside a barn when both doors are shut. Person A sprints with the ladder at a relativistic speed into the barn. A second observer, Person B, is positioned in the barn.

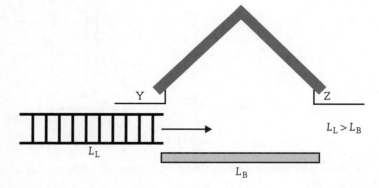

FIGURE 4.5 Situation of the ladder in the barn where $L_L > L_B$ when both the barn and ladder are at rest.

According to Person A, the barn is moving toward them. Hence, it is the barn that experiences length contraction and becomes shorter in length. Hence, the ladder does not fit into the barn.

According to Person B, it is Person A moving at relativistic speeds toward the barn. Hence, the ladder undergoes length contraction and fits inside the barn.

This creates a paradox: how can an event be both true and untrue?

Resolution

The solution to this problem can be developed through the concept of frames of reference. That is, what one observer views of an event (the ladder does not fit) may not be the same as what another observer sees (the ladder does fit), given they are moving relative to each other.

As Person A runs through the barn, they see the ladder unable to fit, whereas Person B – sitting in the middle of the barn - sees the ladder length contract.

Hence, both Person A and Person B are correct in their claims (from their own frame of reference).

Glossary

classical mechanics
the study of motion in accordance with Newton's laws; also known as Newtonian physics

frame of reference
a framework in which the motion of an object is described according to a coordinate system. Frames of reference are observational and can be inertial or non-inertial (accelerating)

inertial frame of reference
one in which Newton's first law applies to a very good approximation, and there is no acceleration. Any departures from the law are negligible; also known as an inertial reference frame

length contraction
length measurements that are shorter in a reference frame that is moving relative to an inertial frame

muon
a subatomic particle formed by cosmic rays in the upper atmosphere; observations of these support special relativity

proper length
length measured in an inertial frame of reference in which the object is stationary

proper time
the time interval between two events occurring at the same place in an inertial reference frame, as measured by an observer in that inertial frame

quantum physics
the science of very small particles for which classical mechanics fails to explain the interactions observed

relative motion
the motion of a moving object according to an observer. When relative motion is being evaluated, one reference frame must always be considered stationary

relativistic effect
when time and space act differently for one object compared to others due to special relativity

relativistic kinetic energy
defined from the rest mass by $E_k = (\gamma - 1)m_0 c^2$

relativistic length
length contraction due to objects moving at very high speeds relative to each other

A+ DIGITAL FLASHCARDS
Revise this topic's key terms and concepts by scanning the QR code or typing the URL into your browser.

https://get.ga/aplus-qce-phys-u34

relativistic mass
also known as relativistically corrected mass. The greater the relative velocity, the greater the relativistic mass: $m = \gamma m_0$

relativistic momentum
momentum of particle due to the relativistic mass at high relative speeds

relativistic time
time dilation observed due to objects moving at very high speeds relative to each other

relativity principle
the laws of physics are the same in all inertial frames of reference

rest energy
defined from the rest mass by $E = m_0 c^2$

rest mass
also known as proper mass; it is the mass as measured when the mass is stationary in an inertial reference frame. Rest or proper mass, m_0, never changes

simultaneity
when two events occur simultaneously in one reference frame but not simultaneously in another reference frame, if the other reference frame is moving relative to the first

special relativity
the physics theory regarding the relationship between space and time, which is not explained by Newtonian or Galilean relativity

time dilation
a longer time measured by an observer outside the reference frame in which the event occurs

Revision summary

Use the following summary of syllabus dot points and key knowledge within Unit 4 Topic 1 to ensure that you have thoroughly reviewed the content. Provide a brief definition or comment for each item to demonstrate your understanding or code them using the traffic light system – Green (all good); Amber (needs some review); Red (priority area to review).

The cognitive verbs have been identified in bold.

Special relativity	
• **describe** an example of natural phenomena that cannot be explained by Newtonian physics, such as the presence of muons in the atmosphere	
• **define** the terms frame of reference and inertial frame of reference	
• **recall** the two postulates of special relativity	
• **recall** that motion can only be measured relative to an observer	
• **explain** the concept of simultaneity	
• **recall** the consequences of the constant speed of light in a vacuum, e.g. time dilation and length contraction	
• **define** the terms time dilation, proper time interval, relativistic time interval, length contraction, proper length, relativistic length, rest mass and relativistic momentum	
• **describe** the phenomena of time dilation and length contraction, including examples of experimental evidence of the phenomena	
• **solve** problems involving time dilations, length contraction and relativistic momentum	
• **recall** the mass–energy equivalence relationship	
• **explain** why no object can travel at the speed of light in a vacuum	
• **explain** paradoxical scenarios such as the twins' paradox, flashlights on a train and the ladder in the barn paradox	

© State of Queensland (QCAA) Physics General Senior Syllabus 2019 CC BY 4.0 https://creativecommons.org/licenses/by/4.0/

Exam practice

Multiple-choice questions

Each multiple-choice question is worth 1 mark.

Solutions start on page 133.

Question 1

Identify which of the following is an example of a natural phenomenon that cannot be explained by Newtonian physics.

A Galilean transformations of relative motion

B Kepler's first law of orbits

C the gravitational acceleration down an inclined plane

D the detection of muons reaching the surface of the Earth

Question 2

Newton's laws of motion are valid in a frame of reference which is:

A accelerating.   **C** inertial.

B at rest.   **D** non-inertial.

Question 3

A spacecraft of known rest length 190 m travels at a speed of 0.80c. The length that the spacecraft appears to be as observed by passengers on board of the spaceship would be:

A 152 m   **C** 114 m

B 237.5 m   **D** 686.75 m

Question 4

Identify which of the following best describes the two postulates of special relativity.

A 1 The laws of physics have the same form in all inertial reference frames.

   2 There is no absolute frame of reference.

B 1 The laws of physics have the same form in all inertial reference frames.

   2 Moving clocks are measured to run slowly.

C 1 A reference frame that moves with constant velocity with respect to an inertial frame is itself also an inertial frame.

   2 All observers in inertial frames of reference agree on the relative speed.

D 1 The laws of physics have the same form in all inertial reference frames.

   2 Light propagates through empty space with a definite speed, c, independent of the speed of the source or the observer.

9780170459174

Question 5

The barn and ladder paradox can be considered from the viewpoint of different observers. The resolution to the paradox is centred on:

A the speed of light changing relative to each observer.

B the fact that all motion is relative.

C faster moving clocks experience dilated time.

D the inertial frames of reference at which an object's length contracts is relative to the observer.

Question 6

Identify which of the following statements best defines the term 'inertial frame of reference'.

A a frame that is stationary relative to net space

B the motion of an object regarding some other moving object

C a non-accelerating frame of reference in which classical laws of physics hold

D an arbitrary set of axes with reference to which the position or motion of something is described or physical laws are formulated

Question 7

Identify which of the following best explains the concept of simultaneity in special relativity.

A the fact that it is impossible to say that two events are in fact simultaneous

B simultaneity is the relation between two events assumed to be happening at the same time

C the concept that two events occurring at the same time is independent of the observer's frame of reference

D the concept that two spatially separated events occur at the same time is not absolute but depends on the observer's reference frame

Question 8

Determine which of the following equations represents a correct rearrangement of the time dilation formula.

> **Hint**
>
> When answering a question asking you to *determine*, remember to follow the QCAA definition.
>
> **determine:** establish, conclude or ascertain after consideration, observation, investigation or calculation; decide or come to a resolution.

A $t_0 = \dfrac{t}{\sqrt{\left(1 - \dfrac{v^2}{c^2}\right)}}$

C $\dfrac{t_o^2}{t^2} = \left(1 - \dfrac{v^2}{c^2}\right)$

B $t = \dfrac{t_0}{\left(1 - \dfrac{v^2}{c^2}\right)}$

D $t = t_0 \sqrt{\left(1 - \dfrac{v^2}{c^2}\right)}$

Question 9

Plutonium-239 is a nuclide that undergoes neutron-induced fission. A typical reaction is included below.

$$^{239}_{94}\text{Pu} + {}^{1}_{0}\text{n} \rightarrow {}^{133}_{55}\text{Cs} + {}^{104}_{46}\text{Pd} + 3\,{}^{1}_{0}\text{n}$$

The exact masses of the nuclides are:

m (Pu-239) = 239.052162 u

m (Cs-133) = 132.905452 u

m (Pd-104) = 103.904030 u

m (neutron) = 1.008665 u

where the atomic mass unit, u, is $1.660\,539 \times 10^{-24}$ kg.

The mass defect, Δm, in this equation was found to be $\Delta m = 3.742 \times 10^{-25}$ kg.

The energy released in joules per atom of plutonium reacted is:

A 3.37×10^{-19} J   C 2.03×10^{16} J

B 3.37×10^{-8} J   D 1.12×10^{-19} J

Question 10

A very fast train travelling at $0.90c$ passes through a very long railway station. A person on board the train (Person A) measures the time interval to go from one end of the station to the other as 1.0 s. What length of time would a person on the platform (Person B) measure the time interval to be?

A   0.44 s   C   1.05 s

B   1.11 s   D   2.29 s

Short response questions

Question 11 (3 marks)

> **Hint**
> When answering a question asking you to *calculate*, remember to follow the QCAA definition.
> **calculate:** determine or find (e.g. a number, answer) by using mathematical processes; obtain a numerical answer showing the relevant stages in the working; ascertain/determine from given facts, figures or information

The Stanford Linear Accelerator (SLAC) can accelerate electrons to over $0.999c$.

a   Calculate the Newtonian momentum and the relativistic momentum of the electron travelling
    at a speed of exactly $0.992c$. The rest mass of an electron is 9.109×10^{-31} kg.   2 marks

b   Determine how many times larger the relativistic momentum value is than the classical
    Newtonian momentum value.   1 mark

Question 12 (5 marks)

a A train passes through a station at a velocity of $25\,\text{m s}^{-1}$. A passenger walks from the rear of the train towards the front at a velocity of $4\,\text{m s}^{-1}$. Apply concepts of relative motion to determine the velocity of the person travelling on the train with respect to:

  i an observer on the station.   1 mark

  ii an observer travelling on the train.   1 mark

b One of the paradoxical scenarios used to explain special relativity is that of the 'ladder in the barn'. Explain the ladder in the barn paradox and how it may be resolved.   3 marks

> **Hint**
>
> When answering a question asking you to *explain*, remember to follow the QCAA definition.
>
> **explain:** make an idea or situation plain or clear by describing it in more detail or revealing relevant facts; give an account; provide additional information

Question 13 (3 marks)

The binary star system of Alpha Centauri A and B is visible to the naked eye from Earth and is located 4.35 light years away. Consider a spacecraft sent there from Earth with a velocity of $0.75c$.

a Determine how many years it would take to reach the binary star system from Earth, as measured by both the observers on Earth as well as observers on the spacecraft.   2 marks

b Calculate the total distance travelled one way, according to the observers on the spacecraft.   1 mark

Question 14 (4 marks)

One of the consequences of the constancy of the speed of light in a vacuum is that of length contraction.

a Explain the concept of length contraction from the point of view of both a stationary observer and a traveller moving with relativistic speeds.   2 marks

b A spacecraft passes Earth with a velocity of $0.80c$. You measure its length to be $1.10 \times 10^2\,\text{m}$. Calculate the length of the spacecraft as it would appear to observers on board the spacecraft.   2 marks

Question 15 (2 marks)

Special relativity suggests that time is not absolute, but rather a relative quantity. Explain this statement and justify your response with an appropriate calculation.

> **Hint**
>
> When answering a question asking you to *justify*, remember to follow the QCAA definition.
>
> **justify:** give reasons or evidence to support an answer, response or conclusion; show or prove how an argument, statement or conclusion is right or reasonable

Question 16 (4 marks)

Determine the momentum of a neutral pi-meson π^0 ($m_0 = 2.48 \times 10^{-28}\,\text{kg}$) travelling at a velocity of $0.994c$ using the:

a relativistic formula.   2 marks

b Newtonian formula.   2 marks

Question 17 (4 marks)

Sienna boards a spacecraft, and flies past Earth at a velocity of 0.85 times the speed of light. Her twin sister, Lucia, remains on Earth. At the instant that Sienna's spacecraft passes Earth, they both start timing the motion of the spacecraft. Sienna stops timing after a period of 45.0 s. Determine how much time has passed on Lucia's stopwatch at this instant.

Question 18 (5 marks)

Einstein conducted a range of thought experiments ('Gendanken'). One of these useful experiments involves a passenger passing a station on a train at 88% of the speed of light, $0.88c$. According to the passenger, the length of the train carriage is 90 m from front to rear. Respond to the following questions using information for this scenario.

> **Hint**
>
> When answering a question asking you to *describe*, remember to follow the QCAA definition.
>
> **describe:** give an account (written or spoken) of a situation, event, pattern or process, or of the characteristics or features of something

a A light in the carriage is switched on. Describe the velocity of the light beam as seen by both the passenger and an observer on the platform.   2 marks

b Calculate the length of the carriage as seen by the stationary observer waiting on the platform. Show your working.   3 marks

Question 19 (3 marks)

Although Einstein had made predictions based on special relativity in the early 20th century, it was not until the later part of the century that precision measurements allowed evidence to be found to support such predictions. Provide one example of a phenomenon that is used to support the theory of special relativity.

Question 20 (4 marks)

a State the two postulates of the theory of special relativity.   2 marks

b Explain why no object may travel at the speed of light, in accordance with the special theory of relativity.   2 marks

Chapter 5
Topic 2: Quantum theory

Topic summary

Quantum theory was developed when classical models, such as the wave model of light, were unable to explain experimental results. The wave–particle duality of light is explored through the study of such experiments and phenomena as Young's double-slit experiment, black-body radiation, atomic models, atomic line spectra, the photoelectric effect and de Broglie wavelength.

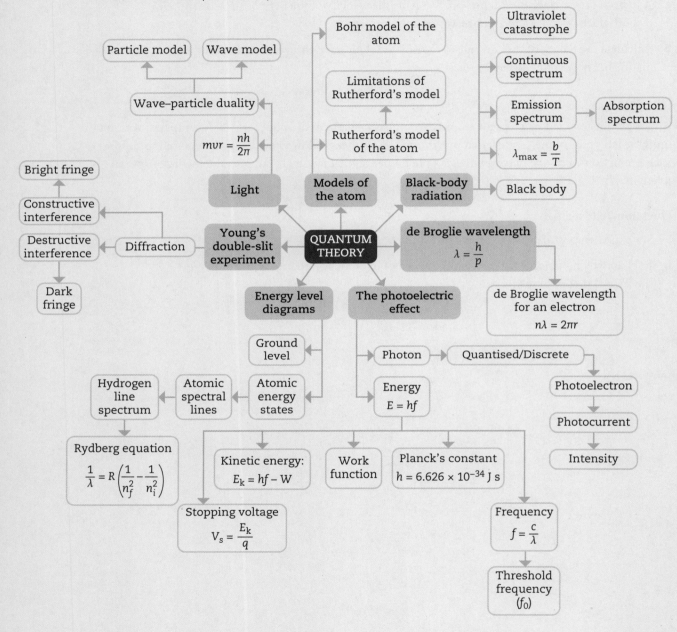

5.1 Young's double-slit experiment

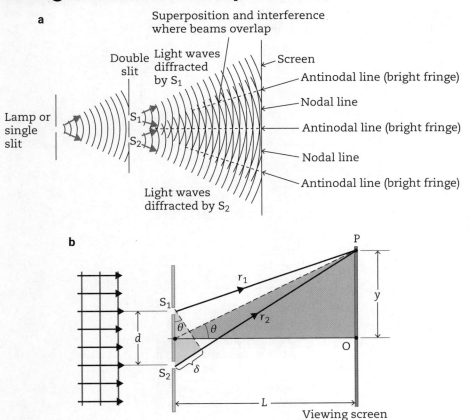

FIGURE 5.1 Young's double-slit experiment. Bright and dark fringes occur on a distant screen when rays of light r_1 and r_2 travelling through nearby slits S_1 and S_2 interfere constructively or destructively. **(a)** Bright fringes (constructive interference) occur along antinodal lines, as crests interact with crests, while dark fringes (destructive interference) occur along nodal lines, as crests interact with troughs. **(b)** Bright fringes occur when the path difference, δ, is equal to a multiple of the wavelength of light, i.e. $\delta = n\lambda$. Dark fringes occur when the path difference, δ, differs by exactly one half of the wavelength and the waves are precisely out of phase, i.e. $\delta = (n - \frac{1}{2})\lambda$.

In the experiment a light source is shone through a pair of narrow, closely spaced slits. The resulting interference pattern can be observed on a distant screen. The interference pattern consists of **constructive interference** (bright fringes) and **destructive interference** (dark fringes).

This is due to the path difference between light waves coming from two different slits.

The interference pattern produced at the screen is due to the two rays travelling different distances to reach the same point. This difference is termed the path difference.

Not only does this experiment show the constructive and destructive interference of light associated with the wave model, it also demonstrates the phenomenon of diffraction that light undergoes – a property of waves. If light followed only the particle model, then the two slits of light would travel directly ahead to make two bright spots on the screen only.

Young's double-slit experiment supports the wave characteristic of light.

5.2 Electromagnetic waves and the universal wave equation

Light can be observed to behave both as a particle and as an electromagnetic wave. Both models are required to completely explain all observed behaviour of matter and energy. Examples of phenomena that support each model are the subject of the Quantum theory topic.

9780170459174

Light can be described as an electromagnetic wave comprising perpendicular magnetic and electric fields that arise as a result of an oscillating electric charge. It behaves just like other waves do – it reflects and refracts and it demonstrates diffraction and interference. This wave model of light is supported by Young's double-slit experiment.

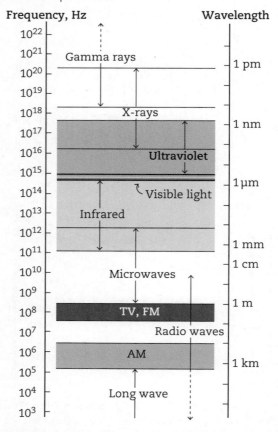

FIGURE 5.2 The electromagnetic spectrum showing all electromagnetic (EM) waves that have been produced or detected, ordered according to their wavelengths and frequencies.

Figure 5.2 shows the full range of electromagnetic waves in the **electromagnetic spectrum** (EMS). This continuous spectrum spans the longer wavelength, lower frequency waves of radio right through to the shorter wavelength, higher frequency (and higher energy) gamma rays.

Solving problems with all forms of electromagnetic radiation, including visible light, require you to apply the universal wave equation relating frequency, f, wavelength, λ, and velocity, v, where the velocity is the speed of light, $c = 3.0 \times 10^8$ m s^{-1} in a vacuum.

$$v = f \times \lambda$$

Hint

Key formula

The wave velocity, v, is directly proportional to the frequency, f, and the wavelength, λ, of the electromagnetic wave.
$v = f \times \lambda$

Worked example

A radio antenna transmits radiowaves with a wavelength of 2.45 m. Calculate the frequency of the electromagnetic wave.

Answer

1 Use the universal wave equation formula.

$v = f \times \lambda$

2 Rearrange for frequency.

$f = \dfrac{v}{\lambda}$

3   Insert known values in the correct units.

$$f = \frac{3.0 \times 10^{-8}}{2.45}$$

4   Calculate the final value.

$$f = 1.22 \times 10^8 \text{ Hz}$$

5.3 Black-body radiation

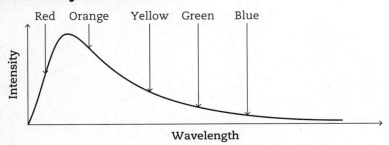

FIGURE 5.3 The intensity of emitted radiation is a function of wavelength for a continuous black-body spectrum.

A **black body** is an ideal surface that completely absorbs all wavelengths of electromagnetic radiation incident on it. It will also be a perfect emitter of electromagnetic radiation at all wavelengths. An example of a black body is a cavity. The radiation that escapes is called **black-body radiation**, and is dependent on the temperature of the cavity.

A black body emission spectrum shows the distribution of wavelengths of radiation and is dependent on temperature (Kelvin).

$$\lambda_{\text{max}} = \frac{b}{T}$$

The position of the peak wavelength is given by Wien's law.

An increase in temperature will shift the peak wavelength to a shorter wavelength with higher energy.

Hint

Key formula

Wien's formula

$$\lambda_{\text{max}} = \frac{b}{T}$$

where:

λ_{max} = peak wavelength (m)

T = absolute temperature (K)

b = Wien's constant, 2.898×10^{-3} m K

5.3.1 The ultraviolet catastrophe

Classical theory predicts that the intensity of radiation should increase as the wavelength decreases, with most energy being released in the ultraviolet region (hence the ultraviolet catastrophe). This was contradicted by experimental data where wave theory could not explain the peak occuring in the intensity of radiation. Wave theory also demonstrated that a black body can release a high amount of energy, which would cause all matter to instantaneously radiate all of the energy it contains, until all matter approaches near absolute zero. The fact that this does not occur was explained with the postulation of Planck's quanta, now known as **photons** (**quantised** light), solving the issue. The phenomenon of black-body radiation is therefore significant in its support for considering light as a particle in the wave–particle duality of light.

5.4   de Broglie wavelength

The concept of de Broglie wavelength, named after Louis de Broglie, suggests that matter exhibits both wave and particle properties, similar to light. This has been supported through experimentation showing the wave nature of electrons. de Broglie claimed that a particle of mass m, moving at a velocity v, would have an associated wavelength, λ. The wavelength λ is known as the de Broglie wavelength.

> **Hint**
>
> **Key formulas**
>
> **The de Broglie wavelength**
>
> $$\lambda = \frac{h}{mv} \text{ or } \lambda = \frac{h}{p}$$
>
> where:
>
> h = Planck constant, 6.626×10^{-34} (J s)
>
> $p = m \times v$ = the momentum of the particle (kg m s^{-1}).
>
> de Broglie's equation also allows us to calculate the momentum of a photon:
>
> **The momentum of a photon**
>
> $$p = \frac{h}{\lambda}$$
>
> where:
>
> h = Planck constant, 6.626×10^{-34} (J s)
>
> p = momentum of the photon (kg m s–1)
>
> λ = de Broglie wavelength of the photon
>
> Further, the de Broglie wavelength for an electron may be determined using its radius of orbit.
>
> **The de Broglie wavelength for an electron**
>
> $n\lambda = 2\pi r$
>
> where:
>
> n = an integer
>
> λ = de Broglie wavelength of the electron (m)
>
> r = radius of the orbit (m)
>
> a 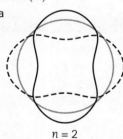   b 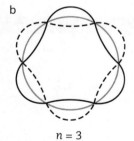   c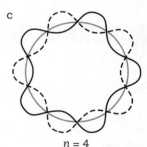
>
> $n = 2$ $n = 3$ $n = 4$
>
> FIGURE 5.4 Electrons as standing waves, showing how energy levels correspond to different standing wave modes: **a** $n = 2$, **b** $n = 3$ and **c** $n = 4$ are in stable orbits.

Worked example

Calculate the de Broglie wavelength of an electron in $n = 2$ with an orbit of radius 4.8 nm.

Answer

$$n\lambda = 2\pi r$$

$$\lambda = \frac{2\pi r}{n}$$

$$\lambda = \frac{2\pi \times 4.8 \times 10^{-9}}{2}$$

$$\lambda = 1.5 \times 10^{-8} \text{ m}$$

> **Hint**
>
> **Key formula**
>
> **Angular momentum of a particle**
>
> $$L = mvr$$
>
> where:
>
> L = angular momentum of the object (kg m^2 s^{-1})
>
> m = mass (kg)
>
> v = velocity (m s^{-1})
>
> r = orbital radius (m)
>
> By saying that L may only have discrete values, then:
>
> $$L = mvr = \frac{nh}{2\pi}$$
>
> where:
>
> h = Planck constant
>
> n = an integer (the quantum number)

Better understandings of the wave nature of electrons supported improved models of the atom.

5.5 The photoelectric effect

5.5.1 Photons and the photoelectric effect

A photon is a **discrete** quanitity of light (**quantum**) of all forms of electromagnetic radiation.

Max Planck deduced that oscillators could only emit and absorb electromagnetic radiation (i.e. light) in packets of specific energies. The amount of energy emitted equals the amount of energy lost when it goes to a lower energy state. One packet of light (photon) has the discrete energy equal to the **frequency** of the light multiplied by **Planck's constant** ($E = hf$).

As the energy required to achieve a higher energy state increases, the probability decreases. Hence, the shorter wavelengths in the black-body emission spectrum exhibit a low frequency. At long wavelengths, there is a high probability of excited states; however, the low energy in each transition leads to low intensity.

Einstein believed that light is a particle (photon) and the flow of photons is a wave, thus upholding the dual nature of light, that is, wave–particle duality.

The photoelectric effect is the ejection of electrons from a polished metal surface by incident light. The incident light must meet a minimum **threshold frequency** for this to occur.

The threshold frequency corresponds to a minimum energy, $E = hf_0$.

The minimum energy corresponds to the **work function**, W, of the metal.

The maximum kinetic energy of the photoelectrons is $E_{k(max)} = hf - W$. This is a statement of conservation of energy.

The photocurrent, which is proportional to the number of photoelectrons, depends on the intensity of the light. The intensity is a measure of the number of incident photons.

$E = hf_0 - W$, hence the gradient of a graph of E_k versus f provides Planck's constant, h.

Hint
Key formula
Photoelectric effect equation
$$E_{k(max)} = hf - hf_0 = E - W$$
where:
$E_{k(max)}$ = kinetic energy of photoelectrons (J)
E = incident photon energy (J)
W = work function (J)

Hint
Key formula
The kinetic energy of a photoelectron
$$E_{k(max)} = qV_s = eV_s$$
where:
$E_{k(max)}$ = kinetic energy (J)
q = charge (C)
V_s = stopping voltage (V)
e = charge on electron (1.60×10^{-19} C)

Hint
Key formula
$$E = hf$$
where:
E = incident photon energy (J)
h = Planck constant, $h = 6.626 \times 10^{-34}$ (J s)
f = frequency of incident light (Hz)

Hint
Key formula
$$W = hf_0$$
where:
W = work function (J)
h = Planck constant, $h = 6.626 \times 10^{-34}$ (J s)
f_0 = threshold frequency (Hz)

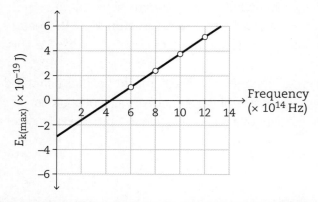

FIGURE 5.5 Graph of kinetic energy (E_k) versus frequency (F) for the photoelectric effect. The gradient represents Planck's constant, h. The x- and y-intercepts represent the threshold frequency, f_0, and work function, W, respectively.

9780170459174

In some instances, light behaves like a particle. This was observed in the photoelectric effect by Heinrich Hertz. He observed that when light is shone on a highly polished metal surface, electrons can be emitted from the surface. The electromagnetic wave model of light could not explain the observations of the photoelectric effect. The experiment showed:

- Electrons were only emitted when the frequency of the light was above a certain threshold, regardless of intensity.
- The number of electrons emitted was proportional to the intensity.

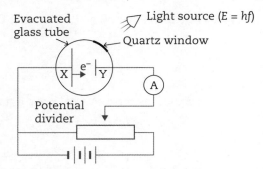

FIGURE 5.6 The photoelectric effect apparatus. An external light source is shone onto a polished metal surface (X). Should the threshold frequency be met, then photoelectrons are ejected from the surface – 1 per photon of light – and a photocurrent is measured by the ammeter, A.

A **photoelectron** is an electron ejected from a metal surface following absorption of a photon of sufficient energy. The emitted photoelectrons are attracted to the positively charged metal plate at Y. An ammeter measures the current produced, termed the **photocurrent**. The size of the current depends on the intensity of the light (number of photons) and not the frequency (as long as the incident frequency, f, is greater than the threshold frequency, f_0, i.e. $f > f_0$).

The maximum kinetic energy can be measured through altering the voltage. The potential difference is reversed to the point where the current becomes zero. At this point the **stopping voltage**, V_{stop}, is found as the potential difference is equal to the maximum energy per unit charge of the electrons. Hence,

$$E_k = q \times V_{stop}$$

Photoelectric effect experimental observations	Unsupported predictions from classical electromagnetic wave theory
A photocurrent occurs only when the frequency of incident light is above the threshold frequency, f_0. f_0 is unique for each metal.	Electrons would be emitted from the surface, regardless of frequency (as long as the intensity is high enough).
The magnitude of the photocurrent is dependent on the intensity and not the frequency.	The photocurrent is dependent on both the intensity and the frequency of the incident light.
No time delay was measurable between the absorption of incident photons and the emission of photoelectrons (at any intensity).	There was a delay in the generation of photocurrent due to time being required for enough energy to be absorbed by the atoms. This delay was predicted to be longer for lower intensities.
The maximum kinetic energy of the photoelectrons depends on the frequency of light ($E_k \alpha f$). The kinetic energy of the photoelectrons is independent of the intensity of incident light.	The kinetic energy is related to the intensity only, and is independent of the frequency.

TABLE 5.1 The photoelectric effect experimental results compared with predictions of classical electromagnetic wave theory.

9780170459174

Worked example

Light of wavelength 225 nm is incident on a polished metal surface with a work function of 5.11 eV.

a What is the value of the work function in joules?

b What is the maximum kinetic energy of the photoelectrons?

c What is the threshold frequency for the metal used?

Answer

1 Convert 5.11 eV to joules using the conversion of 1.60×10^{-19} J per 1.0 electron volts.

$$W = 5.11 \times 1.60 \times 10^{-19} \text{ J}$$

$$W = 8.18 \times 10^{-19} \text{ J}$$

2 Note the photoelectric effect formula as well as the universal wave equation.

$$E_{k(\text{max})} = hf - W$$

$$f = \frac{c}{\lambda}$$

$$E_{k(\text{max})} = \frac{hc}{\lambda} - W$$

$$E_{k(\text{max})} = \frac{6.626 \times 10^{-34} \times 3.0 \times 10^{8}}{225 \times 10^{-9}} - 8.18 \times 10^{-19}$$

$$E_{k(\text{max})} = 8.83 \times 10^{-19} - 8.18 \times 10^{-19}$$

$$E_{k(\text{max})} = 6.55 \times 10^{-20} \text{ J}$$

3 Note the relationship between the threshold frequency and work function.

$$W = hf_0$$

$$f_0 = \frac{W}{h}$$

4 Insert the variable values into the formula.

$$f_0 = \frac{8.18 \times 10^{-19}}{6.626 \times 10^{-34}}$$

$$f_0 = 1.23 \times 10^{15} \text{ Hz}$$

5.6 | Wave–particle duality

Wave–particle duality is the dual nature of light, requiring both wave and particle models to completely explain the phenomena observed.

The photoelectric effect supports viewing light as a particle. A photoelectron is an electron ejected from a metal surface following absorption of a photon of sufficient energy. Light can be described as a quanta of energy that behaves as a particle. It is this behaviour that causes the electrons to be ejected from the metal surface. Hence, the photoelectric effect supports the particle nature of light (and cannot be explained by the wave model of light).

Young's double-slit experiment supports viewing light as a wave. In this experiment a light source is shone through a pair of narrow, closely spaced slits, and there is a resulting interference pattern that is visible on a distant screen. Light thus behaves just like other waves do; that is, light undergoes reflection, refraction, diffraction and interference. Hence, Young's double-slit experiment supports the wave nature of light (and cannot be explained by the particle nature of light).

Einstein believed that light is a particle (photon) and the flow of photons is a wave, thus upholding the dual wave–particle nature of light.

5.7 The Rutherford model of the atom

The Rutherford model of the atom states that the atom is made up of a small, central positively charged nucleus with negatively charged electrons orbiting around it.

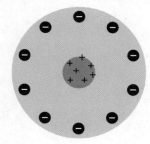

FIGURE 5.7 The Rutherford model of the atom.

Key features of the Rutherford model include:
- Negatively charged electrons orbiting the positively charged nucleus

Limitations of the Rutherford model include:
- The electrons were seen as constantly undergoing centripetal acceleration. This was an issue as accelerating charged particles emit energy. This means the electrons should have been constantly emitting electromagnetic waves.
- Additionally, an electron that was constantly emitting energy would have to gain this energy by spiraling toward the positively charged nucleus. Thus, the atom would collapse.

5.8 The Bohr model of the atom

Bohr proposed that electrons are restricted to certain fixed (quantised) orbits. An electron can jump between these orbits by absorbing or emitting a photon with the appropriate energy, wavelength or frequency.

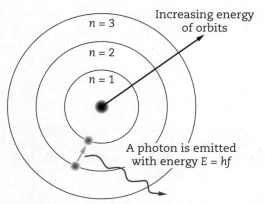

In the Bohr model of the atom the electron can only have certain quantised energy values.

FIGURE 5.8 The Bohr model of the atom. An electron may only exist with certain, discrete energy values.

Key features of the Bohr model include Bohr's postulates.

1 Negative electrons in an atom move in circular orbits about the nucleus under the influence of the electrostatic attraction of the positive nucleus.

2 Only certain orbits are stable, and in these angular momentum takes discrete values. Electrons in these orbits do not emit energy.

3 The radius of the orbit is related to its energy. When an atom absorbs a photon, the electron makes a transition to a higher energy orbit. Atoms emit radiation when an electron goes from an orbit to another orbit with lower energy. The energy released is:

$$E = E_f - E_i = hf$$

Electron orbits are characterised by quantised radii, given by

$$r = \frac{nh}{2\pi m_e v}$$

where:

r = radius (m)

m_e = mass of the electron (kg)

v = velocity (m s^{-1})

h = Planck constant ($h = 6.626 \times 10^{-34}$ J s)

n = an integer

Bohr's model of the atom addressed the limitations of the Rutherford model by explaining the stability of the atom through electrons assuming discrete energy levels. In these stable orbits the electrons do not emit energy, and hence they do not fall in toward the centre of the atom; thus, the law of conservation of energy is upheld.

5.9 Atomic energy states and emission spectra

Bohr's model of the atom explains that an electron may jump between discrete orbits or energy levels. In doing so, a photon is asborbed or emitted with the appropriate frequency, wavelength and energy $E = hf$.

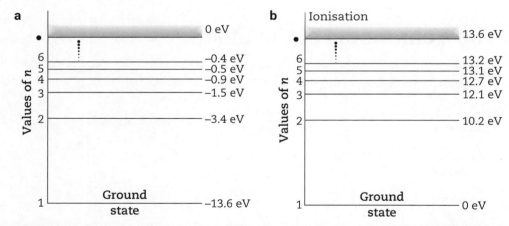

FIGURE 5.9 The energy levels of the hydrogen atom may be represented in two ways (a) with the ionisation energy as zero (0 eV) and (b) with the ground state as zero (0 eV). Either form is acceptable, because in both representations the energy difference between levels is the same. The first representation is generally used as this corresponds to the usual convention for choosing the zero of potential energy.

9780170459174

Bohr's model explains the observed line spectra resulting from such energy level transitions.

A line spectrum is an **emission** or **absorption spectrum** consisting of discrete **atomic spectral lines**, characteristic of the **energy levels** of a particular atom or molecule. The energy released in the form of electromagnetic radiation is found using $E = E_f - E_i = hf$.

Hydrogen produces emissions in the infrared, visible light and ultraviolet sections of the spectrum.

The hydrogen emission and absorption spectra are captured in Figure 5.10. The lines in the emission spectrum of an element coincide with the absorption spectrum at the same wavelength, frequency and hence energy.

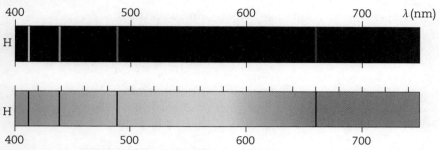

FIGURE 5.10 The emission spectrum and absorption spectrum of hydrogen.

The highest-energy spectral line corresponds to the ionisation energy of electrons.

For hydrogen specifically, the relationship between wavelength and energy level of an electron may be expressed using the Rydberg equation.

Hint
Key formula
Rydberg equation
$$\frac{1}{\lambda} = R\left(\frac{1}{n_f^2} - \frac{1}{n_i^2}\right)$$
where:
R = the Rydberg constant = $1.097 \times 10^7 \, \text{m}^{-1}$
n = an integer
λ = de Broglie wavelength of the photon (m)

Glossary

absorption spectrum
the wavelengths (or frequencies or energies) of radiation absorbed by a material

atomic spectral lines
an emission or absorption spectrum consisting of discrete lines, characteristic of the energy levels of a particular atom or molecule; also called a line spectrum

black body
an object with a perfectly absorbing surface that emits radiation characteristic of its temperature

black-body radiation
the electromagnetic radiation emitted by a black body, with a spectrum characteristic of the temperature of the body

constructive interference
the superposition of waves where crests intersect with crests and and troughs intersect with troughs. It is characterised by antinodes or bright fringes and occurs whenever the path difference is equal to a whole number of wavelengths, $\delta = n\lambda$

continuous spectrum
a spectrum containing radiation of all wavelengths; for example, a rainbow is composed of the various wavelengths of the visible spectrum

destructive interference
the superposition of waves where a crest intersects with a trough, due to incoherent wave sources or sources being half a cycle out of phase. This results in a node where the path difference, δ, is equal to a multiple of one half of the wavelength, $\delta = \left(n - \frac{1}{2}\right)\lambda$

discrete
able to take only specific values, not continuous; for example, a line spectrum is a discrete spectrum

electromagnetic spectrum
the family of electromagnetic radiation – radio waves, microwaves, infrared radiation, visible light, ultraviolet radiation, X-rays and gamma rays – all of which travel at $c = 3.0 \times 10^8\,\mathrm{m\,s^{-1}}$ in a vacuum

energy levels
the allowed energies of a nucleus–electron system; often referred to as electron energy levels, even though they are characteristic of the atom, not of individual electrons

frequency
number of oscillations per time period; $v = f\lambda$ as used in the universal wave equation

A+ DIGITAL FLASHCARDS
Revise this topic's key terms and concepts by scanning the QR code or typing the URL into your browser.
https://get.ga/aplus-qce-phys-u34

line spectrum
an emission or absorption spectrum consisting of discrete lines that are characteristic of the energy levels of a particular atom or molecule

photocurrent
the current formed by electrons ejected from a surface by incident photons

photoelectric effect
the ejection of electrons from a surface by incident photons of sufficient energy

photoelectron
an electron ejected from a metal surface following absorption of a photon of sufficient energy

photon
a particle or quanta of light, having energy $E = hf$

Planck's constant
the constant of proportionality between energy and frequency for photons: $h = 6.626 \times 10^{-34}\,\mathrm{J\,s}$

quantised
existing in discrete amounts, not able to be divided into arbitrarily small amounts

quantum
a discrete unit or amount of some physical property, such as energy, charge, mass or angular momentum

spectrum
the distributed components of light or another wave arranged by frequency (or wavelength)

stopping voltage
the reverse bias voltage required to stop the flow of photoelectrons in a photoelectric effect experiment

threshold frequency, f_0
the minimum frequency of light needed to eject an electron from a metal surface

wave equation
a differential equation that describes wave behaviour; its solutions are wave functions that are typically sinusoids; $v = f\lambda$

wave–particle duality
the dual nature of light, requiring both the wave model and the particle model to completely explain all observed behaviour

work function
the energy required to eject an electron from a metal surface; effectively, it is the ionisation energy for the bulk material

9780170459174

Revision summary

Use the following summary of syllabus dot points and key knowledge within Unit 4 Topic 2 to ensure that you have thoroughly reviewed the content. Provide a brief definition or comment for each item to demonstrate your understanding or code them using the traffic light system – Green (all good); Amber (needs some review); Red (priority area to review)

The cognitive verbs have been identified in bold.

Quantum theory	
• **explain** how Young's double slit experiment provides evidence for the wave model of light	
• **describe** light as an electromagnetic wave produced by an oscillating electric charge that produces mutually perpendicular oscillating electric fields and magnetic fields	
• **explain** the concept of black-body radiation	
• **identify** that black-body radiation provides evidence that electromagnetic radiation is quantised into discrete values	
• **describe** the concept of a photon	
• **solve** problems involving the energy, frequency and wavelength of a photon	
• **describe** the photoelectric effect in terms of the photon	
• **define** the terms *threshold frequency*, *Planck's constant* and *work function*	
• **solve** problems involving the photoelectric effect	
• **recall** that photons exhibit the characteristics of both waves and particles	
• **describe** Rutherford's model of the atom including its limitations	
• **describe** the Bohr model of the atom and how it addresses the limitations of Rutherford's model	
• **explain** how the Bohr model of the hydrogen atom integrates light quanta and atomic energy states to explain the specific wavelengths in the hydrogen line spectrum	

›>

››	• **solve** problems involving the line spectra of simple atoms using atomic energy states or atomic energy level diagrams	
	• **describe** wave–particle duality of light by identifying evidence that supports the wave characteristics of light and evidence that supports the particle characteristics of light	
	• Mandatory practical: **Conduct** an experiment (or use a simulation) to investigate the photoelectric effect	

© State of Queensland (QCAA) Physics General Senior Syllabus 2019 CC BY 4.0 https://creativecommons.org/licenses/by/4.0/

Exam practice

Multiple-choice questions

Each multiple-choice question is worth 1 mark.

Solutions start on page 136.

Question 1

The displacement law that relates the emissions of a black body (an ideal substance that emits and absorbs all frequencies of light) and its temperature was proposed by:

A Einstein.

B Faraday.

C Maxwell.

D Wein.

Question 2

The calculated maximum energy, in J and eV, of a quantum (photon) of green light electromagnetic radiation of frequency 5.45×10^{14} Hz is:

> **Hint**
>
> When answering a question asking you to *calculate*, remember to follow the QCAA definition.
>
> **calculate:** determine or find (e.g. a number, answer) by using mathematical processes; obtain a numerical answer showing the relevant stages in the working; ascertain/determine from given facts, figures or information

A 3.61×10^{-19} J, 0.08 eV

B 3.61×10^{-19} J, 2.26 eV

C 1.22×10^{-20} J, 0.08 eV

D 1.22×10^{-20} J, 2.26 eV

Question 3

Rutherford described the structure of the atom in terms of an electron orbiting a small, dense, positively charged nucleus. Although his planetary model of the atom fit with most experimental data, there was a main limitation to Rutherford's atomic theory. This main limitation was:

A the model could not explain the centralised nucleus of the atom.

B the model did not identify the difference in mass of positive and negative charges.

C the protons in this model were restricted to certain orbits characterised by discrete energies.

D the charged electron in this model experiences circular motion and centripetal acceleration and hence should continuously radiate electromagnetic energy.

Question 4

Identify which of the following statements best represents the photoelectric effect.

A the ionising of metal atoms by photons

B the phenomenon discovered whereby light may be considered a wave

C metals absorbing photons with frequencies equal to or greater than the threshold frequency

D when a photon of incident light provides enough energy to eject an electron from a metal's surface

Question 5

Young's double-slit experiment shone monochromatic light through two slits onto a viewing screen to form interference patterns. This experiment provides evidence to support:

A the photoelectric effect.

B the wave behaviour of light.

C electron-positron annihilation.

D the particle behaviour of light.

Question 6

A beam of light has a wavelength of 4.50×10^{-8} m.

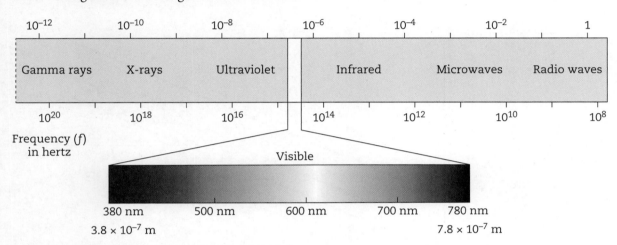

Its wavelength, frequency and spectral region are:

A 45 nm, 6.67×10^{15} Hz, ultraviolet

B 450 nm, 6.67×10^{15} Hz, visible

C 45 nm, 13.5 Hz, ultraviolet

D 450 nm, 13.5 Hz, visible

Question 7

A copper surface requires photons with a frequency of at least 1.10×10^{15} Hz to release photoelectrons. Determine the maximum velocity of ejected photoelectrons released when an incident light of wavelength 380 nm hits the surface.

A 5.23×10^{5} m/s

B 2.06×10^{10} m/s

C 6.73×10^{5} m/s

D 4.52×10^{5} m/s

Question 8

Light of 233 nm is incident upon a metal surface with a work function of $W = 1.69 \times 10^{-19}$ J.

Which of the following statements best describes the possible outcome?

A No photoelectrons are ejected.

B Photoelectrons with a maximum kinetic energy of 1.69×10^{-19} J are ejected.

C Photoelectrons with a maximum kinetic energy of 6.85×10^{-19} J are ejected.

D Photoelectrons with a maximum kinetic energy of 9.85×10^{-19} J are ejected.

Question 9

The intensity of the incident light on the metal surface of a photocell is increased. This increase also results in an increase of:

A the kinetic energy of the electrons.

B the photocurrent.

C the threshold frequency.

D the work function.

Question 10

A graph of the maximum kinetic energy of photoelectrons for different metals is shown below.

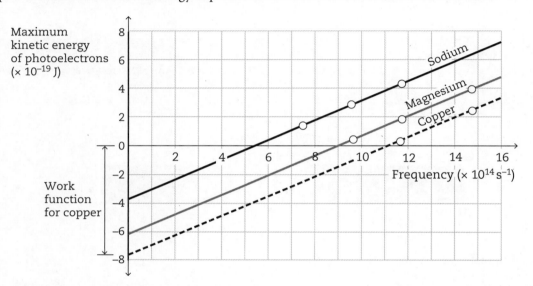

Determine the characteristic that is common to each metal.

A the work function

B the threshold frequency

C the ratio of the maximum kinetic energy and frequency of incident light

D the maximum kinetic energy with an incident light frequency of $8 \times 10^{14}\,\text{s}^{-1}$

Short response questions

Question 11 (3 marks)

The Bohr model of the hydrogen atom integrates the quantisation of light and atomic energy states. Explain how this model relates specific wavelengths of light to the energy levels of hydrogen line spectrum. Use the energy level diagram for hydrogen and an example calculation of a wavelength of emitted light from the transition from level 5 to level 2 to support your explanation.

> **Hint**
>
> When answering a question asking you to *explain*, remember to follow the QCAA definition.
>
> **explain:** make an idea or situation plain or clear by describing it in more detail or revealing relevant facts; give an account; provide additional information

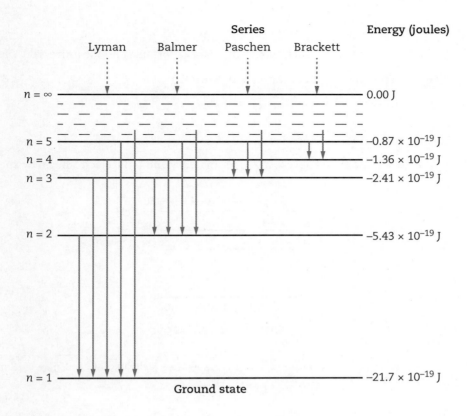

Question 12 (3 marks)

a Construct a line on the energy level diagram to represent a transition where an electron absorbs a photon of 2.9×10^{-19} J. Show the relevant calculation for this transition.   2 marks

Energy (joules)

$n = 4$ ——————— 0.00 J

$n = 3$ ——————— -2.50×10^{-19} J

$n = 2$ ——————— -5.97×10^{-19} J

$n = 1$ ——————— -8.87×10^{-19} J

b Use the figure above to determine the maximum frequency of light that may be emitted from the transitions of the mercury atom shown.   1 mark

Question 13 (3 marks)

Describe the wave–particle duality of light. Use one piece of evidence that supports the wave characteristic of light and a second piece of evidence that supports the particle characteristic of light.

> **Hint**
> When answering a question asking you to *describe*, remember to follow the QCAA definition.
>
> **describe:** give an account (written or spoken) of a situation, event, pattern or process, or of the characteristics or features of something

Question 14 (3 marks)

Determine the following values using Planck's constant, Planck's formula and the universal wave equation.

a   The energy, in eV, of a photon of light of wavelength 650 nm.   2 marks

b   The wavelength, in nm, of an X-ray with energy 2.9 keV.   1 mark

Question 15 (4 marks)

The diagrams show **a** the atomic energy levels, in eV, for mercury and **b** the electromagnetic spectrum.

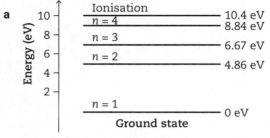

Energy levels for Hg

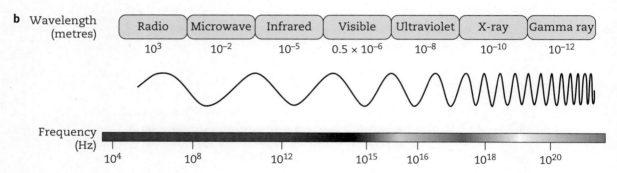

Demonstrate that when an electron drops from its first excitation energy level to the ground state, an ultraviolet photon is emitted.

> **Hint**
>
> When answering a question asking you to *demonstrate*, remember to follow the QCAA definition.
>
> **demonstrate:** prove or make clear by argument, reasoning or evidence, illustrating with practical example; show by example; give a practical exhibition

Question 16 (2 marks)

Black-body radiation provides evidence for the quantisation of electromagnetic radiation. Explain this phenomenon using key features of the model.

Question 17 (2 marks)

A particle measured to be travelling with a velocity of $2.10 \times 10^4 \, \text{m s}^{-1}$ has an equivalent energy to a photon of light of frequency 3.5×10^{14} Hz. Deduce the mass of the particle.

> **Hint**
>
> When answering a question asking you to *deduce*, remember to follow the QCAA definition.
>
> **deduce:** reach a conclusion that is necessarily true, provided a given set of assumptions is true; arrive at, reach or draw a logical conclusion from reasoning and the information given

Question 18 (6 marks)

A magnesium metal cathode with a work function of 3.70 eV is illuminated with light of wavelength 250 nm.

> **Hint**
>
> When answering a question asking you to *determine*, remember to follow the QCAA definition.
>
> **determine:** establish, conclude or ascertain after consideration, observation, investigation or calculation; decide or come to a resolution.

a Determine whether the photoelectric effect may be demonstrated in this instance. Justify your response.   3 marks

b Determine the maximum kinetic energy, E_k, of any ejected photoelectrons.   1 mark

c Calculate the maximum velocity of a photoelectron, if ejected. Use the mass of an electron, $m_e = 9.1 \times 10^{-31}$ kg.   2 marks

Question 19 (5 marks)

a Identify three key features of the photoelectric effect.   3 marks

b Describe one example of how the photoelectric effect provides evidence of the quantisation of photons.   2 marks

Question 20 (4 marks)

The Bohr model of the atom addressed several limitations of the Rutherford model. Identify the key limitations of the Rutherford model and explain how Bohr's model of the atom overcame them.

Chapter 6
Topic 3: The Standard Model

Topic summary

The Standard Model of Particle Physics provides our current understanding of the universe on the smallest of scales. Discoveries in particle physics have led to theories of an underlying structure and behaviour of particles that describe and predict both elementary particles and their possible interactions. The Standard Model is explored through understanding the fundamental forces as well as the particles called leptons, hadrons, mesons, baryons and quarks.

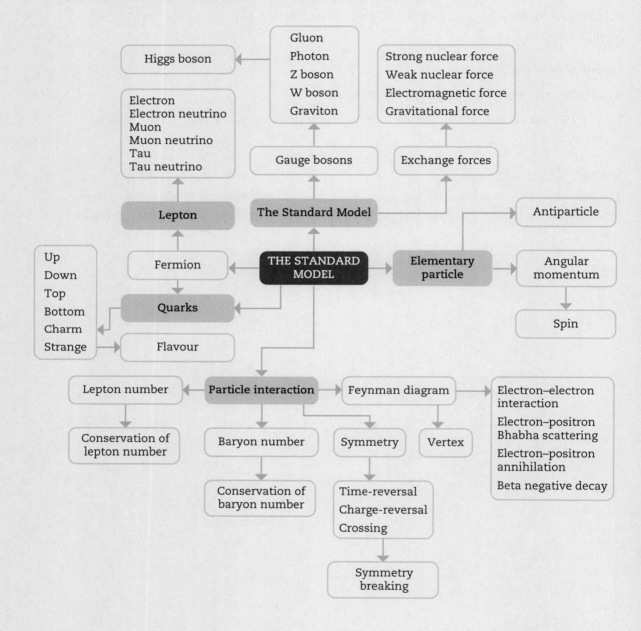

6.1 The Standard Model

The elementary particles are the building blocks of the universe and are grouped according to specific properties. At a fundamental level, these particles have interactions that can be represented theoretically across space and time. Conservation of the elementary particles as well as the symmetry in particle interactions allow scientists to determine further potential particles and interactions. The Standard Model of Particle Physics describes the electromagnetic, weak, and strong interactions and is used to classify all elementary particles known to humankind.

6.1.1 Elementary particles and antiparticles

An **elementary particle**, also known as a fundamental particle, is a subatomic particle with no known substructure.

An **antiparticle** is a particle with the same mass and the opposite charge and/or **spin** to a corresponding particle (e.g. **positron** and electron)

Feynman diagrams are models used to show exchange particles and **exchange forces** over time in space, when particles come into close proximity to each other. These will be addressed later in this chapter, with particles being drawn in the direction of time and their antiparticles being shown in the opposite direction to time.

Quarks

Quarks are subatomic particles that are governed by the strong nuclear force.

Hadrons are subatomic composite particles comprised of quarks. There are six quarks in the Standard Model.

- Up (u)
- Down (d)
- Charm (c)
- Strange (s)
- Top (t)
- Bottom (b)

TABLE 6.1 The six quarks are easy to remember in their somewhat oppositely named pairs.

u	c	t
d	s	b

Hint

It is easy to remember the six quarks and their antiquarks by recalling the relevant section of the table of the Standard Model. Redraw this until you recall all the elementary particles.

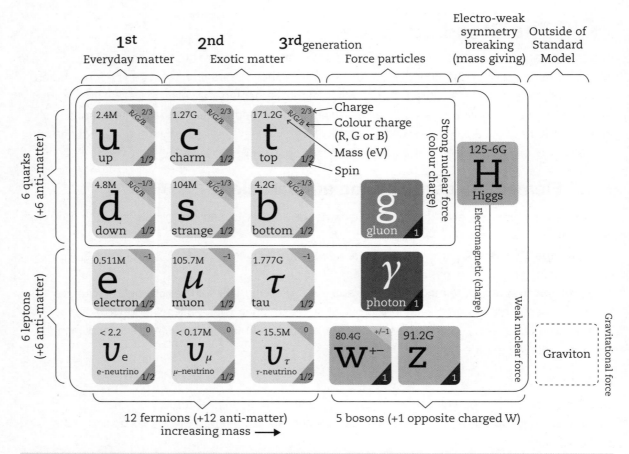

FIGURE 6.1 The 'periodic table' of the elementary particles of the Standard Model of Particle Physics.

Some further particles and their properties are listed in Table 6.2. The various ways of classifying these particles will be explored in the following sections. Note that masses are given in the units of MeV c^{-2}. (Remember that mass and energy are equivalent, related through $E = mc^2$ and 1.0 MeV $c^{-2} = 1.78 \times 10^{-30}$ kg.)

TABLE 6.2 Elementary particles of the Standard Model of Particle Physics and their properties (baryon number, lepton number, lifetime and spin).

Particle name	Symbol	Anti-particle	Mass (MeV c⁻²)	B	Le	Lμ	Lτ	Lifetime (s)	Spin
Leptons									
Electron	e−	e+	0.511	0	+1	0	0	Stable	$\frac{1}{2}$
Electron-neutrino	v_e	$\bar{v}_e$	<7 eV c⁻²	0	+1	0	0	Stable	$\frac{1}{2}$
Muon	μ⁻	μ⁺	105.7	0	0	+1	0	2.20×10^{-6}	$\frac{1}{2}$
Muon-neutrino	v_μ	$\bar{v}$	<0.3	0	0	+1	0	Stable	$\frac{1}{2}$
Tau	τ	τ⁺	1784	0	0	0	+1	<4 × 10⁻¹³	$\frac{1}{2}$
Tau-neutrino	v_τ	$\bar{v}_\tau$	<30	0	0	0	+1	Stable	$\frac{1}{2}$

Particle name	Symbol	Anti-particle	Mass (MeV c⁻²)	B	Le	Lμ	Lτ	Lifetime (s)	Spin
Hadrons – mesons									
Pion	π^+	π^-	139.6	0	0	0	0	2.60×10^{-8}	0
	π^0	Self	135.0	0	0	0	0	0.83×10^{-16}	0
Kaon	K^+	K^-	493.7	0	0	0	0	1.24×10^{-8}	0
	$\overline{K}^0_S$	$\overline{K}^0_S$	497.7	0	0	0	0	0.89×10^{-10}	0
	$\overline{K}^0_L$	$\overline{K}^0_L$	497.7	0	0		0	5.2×10^{-8}	0
Eta	η	Self	548.8	0	0	0	0	$<10^{-8}$	0
	η'	Self	958	0	0	0	0	2.2×10^{-10}	0
Hadrons – baryons									
Proton	p	$\bar{p}$	938.3	+1	0	0	0	Stable	$\frac{1}{2}$
Neutron	n	$\bar{n}$	939.6	+1	0	0	0	614	$\frac{1}{2}$
Lambda	Λ^0	$\overline{\Lambda}^0$	1115.6	+1	0	0	0	2.6×10^{-10}	$\frac{1}{2}$
Sigma	Σ^+	$\overline{\Sigma}^-$	1189.4	+1	0	0	0	0.80×10^{-10}	$\frac{1}{2}$
	Σ^0	$\overline{\Sigma}^0$	1192.5	+1	0	0	0	63×10^{-20}	$\frac{1}{2}$
	Σ^-	$\overline{\Sigma}^+$	>1197.3	+1	0	0	0	1.5×10^{-10}	$\frac{1}{2}$
Delta	Δ^{++}	$\overline{\Delta}^{--}$	1230	+1	0	0	0	6×10^{-24}	$\frac{3}{2}$

Baryons and mesons

Hadrons are categorised into two families: baryons and mesons.

Baryons are composite subatomic particles made up of three quarks (e.g. protons, neutrons).

Mesons are subatomic particles composed of one quark and one antiquark (e.g. pions).

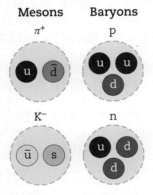

FIGURE 6.2 The quark composition of two mesons and two baryons. Mesons consist of a quark–antiquark pair, while baryons consist of three quarks.

Leptons

Leptons are particles that are governed by the weak nuclear force and, since they have charge, are also influenced by electromagnetism. There are six leptons in the Standard Model:

- electron (e)
- electron neutrino (v_e)
- muon (**μ**)
- muon neutrino (v_μ)
- tau (**τ**)
- tau neutrino (v_τ)

TABLE 6.3 The six leptons are easy to remember in pairs with their accompanying neutrinos.

e	μ	τ
v_e	v_μ	v_μ

Hint

It is easy to remember the six leptons and their antileptons by recalling the relevant section of the table of the Standard Model. Redraw this until you recall all the elementary particles.

Gauge bosons

Gauge bosons are carriers or **exchange particles** that govern particle ineraction and mediation of the four fundamental forces. **Bosons** do not obey the **Pauli exclusion principle**. There are four fundamental forces and four gauge bosons in the Standard Model. There is a theoretical fifth gauge boson that mediates the gravitational force, the **graviton**; however, its existence has not yet been proven.

The four gauge bosons are:

- the gluon
- the photon
- the Z boson
- the W⁺ and W⁻ bosons

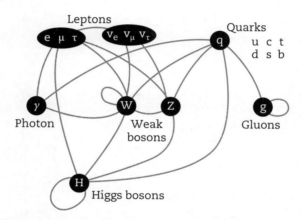

FIGURE 6.3 The types of particles (fermions and bosons) of the Standard Model of Particle Physics.

6.1.2 Fundamental forces and gauge bosons

The strong nuclear force is the attractive force between quarks, holding them together (think: proton–proton repulsion in the nucleus). It is mediated by the gluon.

TABLE 6.4 The fundamental forces and their relative strengths and interactions.

Force	Gauge boson	Strength	Acts on
Strong	Gluon	1	Interacts with quarks
Electromagnetic	Photon	10^{-2}	Interacts with charged particles
Weak	W and Z	10^{-6}	Interacts with neutrinos and beta decay
Gravity	Graviton	10^{-39}	Interacts with mass-containing systems

The weak nuclear force is the fundamental force responsible for radioactive decay. It is mediated by the **W and Z bosons**.

Electrons and other charged particles interact via electric and magnetic fields, and this interaction is governed by the electromagnetic force. It is mediated by the photon gauge boson.

Experiencing fundamental forces

The strong nuclear force is experienced by quarks.

The weak nuclear force is experienced by both quarks and electrons.

The electromagnetic force is experienced by charged leptons (e.g. electrons).

The gravitational force is experienced by all particles with mass.

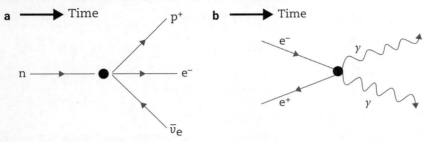

FIGURE 6.4 a A simple reaction diagram showing a neutron decaying to a proton, electron and an electron antineutrino. **b** Positron–electron annihilation results in two photons being produced.

6.2 Conservation of lepton number and baryon number

6.2.1 Lepton number

Leptons are particles that do not interact by means of the strong nuclear force, but do interact via the gravitational, electromagnetic and weak nuclear forces.

Leptons are divided into three lepton families: the electron and its neutrino, the muon and its neutrino and the tau and its neutrino.

Electrons and their neutrinos have electron number +1, positrons and their antineutrinos have electron number −1, and all other particles have electron number 0. Both the muon family and the tau family follow the same pattern.

Hence, there are three different **lepton numbers**: L = +1 for all real leptons, L = −1 for all antileptons and L = 0 for all other non-lepton particles.

Specifically, by family:

L_e (electron) = +1 (for electrons and electron neutrinos), L_e (positron) = −1 (for positrons and electron antineutrinos).

L_μ (muon) = +1 (for **muons** and muon neutrinos), L_μ (antimuon) = −1 (for antimuons and muon antineutrinos), and

L_τ (tau) = +1 (for taus and tau neutrinos), L_τ (antitau) = −1 (for antitaus and tau antineutrinos).

For all other particles the lepton number = 0, for example, a muon has a L_τ = 0, a neutron has a L_e = 0.

The lepton numbers are conserved in families during reactions.

6.2.2 Baryon number

Baryons are composite subatomic particles made up of three quarks.

Every particle can be assigned a **baryon number**:

B = +1 for all real baryons, B = −1 for all antibaryons, B = 0 for all other non-baryon particles.

For example: B (proton) = +1, B (antiproton) = −1, B (pion) = 0.

Baryon numbers are used to describe **annihilation** and the creation of particles. Whenever a baryon is created, an antibaryon is also created.

TABLE 6.5 The baryon numbers associated with the quarks

Quark	Symbol	Charge	Baryon number
up	u	+2/3 e	1/3
down	d	−1/3 e	1/3
charm	c	+2/3 e	1/3
strange	s	−1/3 e	1/3
top	t	+2/3 e	1/3
bottom	b	−1/3 e	1/3

The baryon numbers are also conserved during a reaction.

For example, a proton–neutron pair production reaction conserves baryon number, B.

$$p + n \rightarrow p + n + \bar{p} + p$$

$$\text{Baryon number, B} : 1 + 1 \rightarrow 1 + 1 + (-1) + 1$$

The baryon number is defined by the formula, $B = \frac{1}{3}(n_q - n_{\bar{q}})$, where n_q is the number of quarks and $n_{\bar{q}}$ is the number of antiquarks. This definition may be used to determine the baryon number of any particle, for example.

TABLE 6.6 Baryon numbers may be determined using the formula $B = \frac{1}{3}(n_q - n_{\bar{q}})$

Particle	Quark composition	Baryon number
Proton	uud	$B = \frac{1}{3}(3-0) = 1$
Neutron	udd	1
antiproton	$\bar{u}\bar{u}\bar{d}$	−1
antineutron	$\bar{u}\bar{d}\bar{d}$	−1

6.2.3 Conservation laws

The law of **conservation of lepton number** and the law of **conservation of baryon number** must be adhered to in any reaction or decay. This means that the reactants and products of any given interaction must have the same baryon and lepton numbers.

Consider the decay of a neutron into a proton:

$$n \rightarrow p + e^- + \bar{v}_e$$

On the left side, there is one neutron. Therefore the baryon number is 1, and the electron lepton number is 0. On the right side, there is a proton, electron and an electron antineutrino. Hence, the baryon number is 1 and the electron lepton number is 0, as $L_e = 1 - 1 = 0$.

If the lepton or baryon number is not conserved in a reaction, the reaction is not possible.

In fact, if any of the conservation laws are violated then the decay is not possible. This includes the conservation laws of energy, linear momentum, angular momentum (spin), charge, lepton number and baryon number.

6.3 Particle interactions

If lepton number is not conserved in a reaction, it means that the process is not possible. In the instance where a neutrino with a given lepton association (e, μ or τ) changes to another neutrino type (e, μ or τ), lepton number is still conserved. This is a phenomena called **neutrino oscillation**.

6.3.1 Feynman diagrams

Feynman diagrams are a type of model used to show exchange particles and exchange forces over time in space, when particles come into close proximity to each other.

The following conventions are used:

* Time is measured in the positive *x* direction
* Space is measured in the positive *y* direction
* Particles are represented as straight line arrows in the direction of time, with a letter indicating the type.
* Particles are shown in the direction of time; antiparticles are shown in the opposite direction to time.
* Gauge bosons are represented by wiggly lines and a letter indicating the boson represented. The gluon can also be represented by a string of loops.
* The point at which a particle interacts is called a vertex. A vertex always has three lines attached to it.

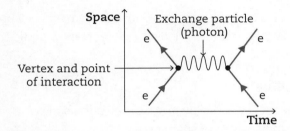

FIGURE 6.5 Labelled Feynman diagram. The exchange particle (a photon in this case) happens at the same instant the two electrons come close together. This exchange causes the electrostatic repulsion of the electrons. Note: The arrows do not note a trajectory; they indicate that the electron is a particle moving in time.

The syllabus requires students to be familiar with three types of Feynman diagrams: electron–electron interactions, electron–positron interactions, and a neutron decaying into a proton.

6.3.2 Electron–electron interactions

When an electron interacts with another electron, they scatter off unchanged in various directions. As the electrons come closer together, they interact by repelling as a result of electrostatic repulsion, but do not change into something else. Thus, this interaction is mediated by the electromagnetic force.

Its equation is as follows:

$e^- + e^- \rightarrow e^- + e^-$

This interaction can be described in a Feynman diagram:

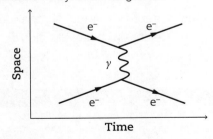

FIGURE 6.6 Feynman diagram of an electron–electron interaction.

The force carrier (the photon) is created, then acts between them and disappears. It is called a 'virtual' photon as it only exists as long as the interaction takes place.

The wiggly line representing the gauge boson is vertical, indicating that there is not a period of time where the particles are converted into energy.

6.3.3 Electron–positron interactions

When an electron and a positron interact, there are two possible outcomes: Bhabha scattering, or electron–positron annihilation.

Bhabha scattering

At the moment of interaction the electron and positron exchange a virtual photon (γ) and scatter off each other with only their velocities changed.

This interaction can be described using the following Feynman diagram:

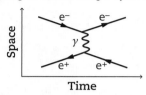

FIGURE 6.7 Feynman diagram of an electron–positron scattering interaction.

The force carrier (the photon) is created, then acts between them and disappears.

The gauge boson is represented vertically, indicating that no new particles are created, and only their velocities are changed.

Electron–positron annihilation

Another process that can happen is annihilation. As the electron and positron get closer, they annihilate (wipe out) each other instead of scattering.

They form a virtual photon, which lives for a very short amount of time before forming a new electron–positron pair. This is a great example of energy being expressed as matter.

This interaction can be described using the following Feynman diagram:

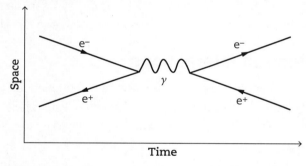

FIGURE 6.8 Feynman diagram of an electron–positron annihilation interaction.

In an electron–positron annihilation interaction a virtual photon is created, which lives for a short period of time before forming a new electron–positron pair. This gauge boson is represented horizontally, indicating there is a period of time where the particles are converted to the virtual photon before new particles are formed from the old ones.

> **Hint**
>
> To tell the difference between an electron–positron scattering and an electron–positron annihilation Feynman diagram, look at the gauge boson (the wiggly line). In an annihilation interaction, the gauge boson is horizontal because, as the name indicates, there is a period of time where the electron and positron do not exist. In contrast, in a scattering interaction the gauge boson is vertical because there is never a period of time where the electron and positron do not exist.

6.3.4 Neutron to proton transmutation (beta negative decay)

Beta negative decay is the decay of a neutron into a proton. This process occurs when an atom has too many neutrons for the number of protons, and thus is unstable.

   The atom uses some of its excess energy by undergoing beta decay, in which one of the neutrons turns into a proton. In doing so this energy is used to form a temporary W⁻ boson that disappears by forming an electron and an electron antineutrino.

$$n \rightarrow p + e^- + \bar{v}_e$$

FIGURE 6.9 Beta negative decay equation.

This interaction can be described using the following Feynmn diagrams:

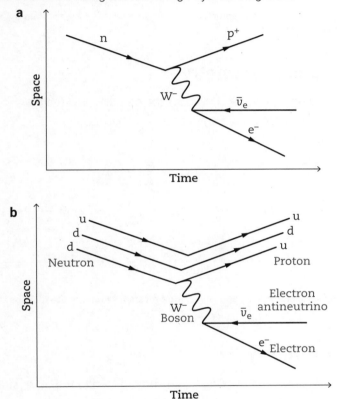

FIGURE 6.10 a Feynman diagram of a beta negative decay interaction b Feynman diagram of a beta negative decay interaction including quark composition.

The second diagram is more useful than just showing a proton and neutron as it shows the quarks. One of the quark changes **flavour** from down to up, making it a proton.

> **Hint**
> Think: beta negative decay, therefore the W boson is negative (W⁻).

6.4 Symmetry in particle interactions

There are three **symmetries** in particular that give us the means of predicting possible particle interactions. These are **time-reversal symmetry**, **charge-reversal symmetry**, and **crossing symmetry**. If we apply any of these symmetries to an allowed reaction, then the resulting reaction is also allowed.

6.4.1 Time-reversal symmetry

Time-reversal symmetry involves flipping the order of the reaction such that it occurs in reverse order. As this effectively flips that time axis on the Feynman diagram, this form of symmetry is called **time reversal**. This is a useful tool to determine other possible interactions.

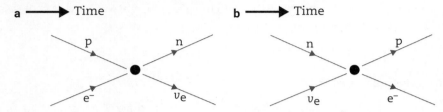

FIGURE 6.11 Time reversal applied to electron capture. a An electron is captured by a proton. **b** A neutron reacts with a neutrino to form a proton and electron.

6.4.2 Charge-reversal symmetry

Charge-reversal symmetry says that if the charges on all particles in a reaction are reversed, the new reaction is also possible and will not violate any conservation laws.

i.e. $\mu^+ \rightarrow e^+ + \nu_e + \bar{\nu}_\mu$ and $\mu^- \rightarrow e^- + \bar{\nu}_e + \nu_\mu$

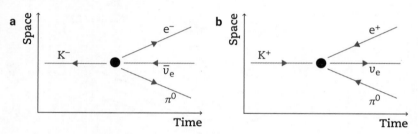

FIGURE 6.12 a Decay of a negative kaon **b** once applying charge-reversal symmetry, the reaction turns into positive kaon decay.

6.4.3 Crossing symmetry

Crossing symmetry involves one particle being taken and crossed to the other side of the reaction, and converted to its antiparticle. The crossing symmetry interaction is also deemed valid.

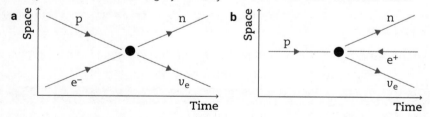

FIGURE 6.13 a $p + e^- \rightarrow n + \nu_e$ **b** New reaction after crossing the electron: $p \rightarrow n + \nu_e + e^+$. **This is an observed reaction.**

Symmetry breaking

The rules of symmetry (time-reversal, charge-reversal and crossing) show that these alternative particle interactions may occur, even if they are unlikely. In some cases, the new particle interaction that is derived from the base particle interaction does not occur. This is due to **symmetry breaking**, which occurs when there is a change in the behaviour of a physical system or the laws of physics that then prevents a translation, reflection or rotation in time or space from taking place.

Glossary

annihilation
the destructive process resulting when a particle and its antiparticle meet

antiparticle
a particle with the same mass and opposite charge and/or spin to a corresponding particle

baryon number
quantum number associated with each baryon, antibaryon and non-baryonic particle

baryons
a family of heavy subatomic particles, such as neutrons and protons, which contain composite structures made up of three quarks

bosons
particles with integer spin ($s = 0, 1, 2, \ldots$); these particles do not obey the Pauli exclusion principle

charge-reversal symmetry
if all particles in an allowed reaction are replaced with their antiparticles (which have opposite charge), the new reaction is also allowed under known conservation laws

conservation of baryon number
whenever a nuclear reaction or decay occurs, the sum of the baryon numbers before the process must equal the sum of the baryon numbers after the process

conservation of lepton number
each of the lepton numbers L_e, L_μ and L_τ is a conserved quantity

crossing symmetry
if a particle in an allowed reaction is crossed to the other side of the reaction and replaced with its antiparticle, the new reaction is also allowed under known conservation principles, provided enough energy is available

elementary particle
a particle whose substructure is unknown

exchange force
strong, electromagnetic, weak or gravitational force associated with the exchange particles gluons, photons, W and Z bosons and gravitons respectively; for example, an exchange of photons between electrons is mediated by the electromagnetic force

exchange particle
a particle carrying force which is responsible for behaviour during other particle interactions. Sometimes exchange particles are the result of a particle interaction, such as the case with electron–positron annihilation.

A+ DIGITAL FLASHCARDS
Revise this topic's key terms and concepts by scanning the QR code or typing the URL into your browser.

https://get.ga/aplus-qce-phys-u34

fermions
particles with half-integer spin

Feynman diagram
a diagram that models exchange particles and exchange forces over time in space, when particles come into close proximity to each other

flavours
the six classifications of quark types: up, down, strange, charm, top and bottom

gauge boson
force-carrying particles that mediate particle interactions through the four fundamental forces

gluon
the gauge boson that mediates the strong force

graviton
the hypothetical gauge boson of the gravitational force

hadron
a family of elementary particles with a large mass consisting of mesons and baryons

lepton number
quantum number associated with each lepton, antilepton and non-leptonic particle

leptons
a family of elementary particles that includes electrons, taus, muons, their neutrinos and all of their antiparticles

mesons
a family of heavy subatomic particles that contain composite structures made up of one quark and an antiquark

muon
a particle formed by cosmic rays in the upper atmosphere

neutrino oscillation
a phenomenon in which a neutrino with a given lepton association (e, μ or τ) can later be measured to have switched to another neutrino type (e, μ or τ); lepton number is still conserved in this instance

Pauli exclusion principle
quantum mechanical principle that two fermions in the same quantum system cannot have identical sets of quantum numbers; e.g. no two electrons can be in the same shell or orbital around an atom and have the same energy

positron
the antiparticle of an electron with charge +e and mass m_e

quark
a type of elementary particle (along with leptons and gauge bosons)

spin
a quantum property of particles that results from them having their own magnetic moment, and therefore, magnetic field

symmetry
the invariance of physical laws under transformations such as translation, reflection or rotation in time or space

symmetry breaking
a change in the behaviour of a physical system or the laws of physics that govern its behaviour when a symmetry operation such as a translation, reflection or rotation in time or space takes place

time reversal
when reactions are reversed in time

time-reversal symmetry
when an allowed reaction is written such that it runs in the opposite direction in time; the new reaction is also allowed in that it does not break any of the known conservation laws

W and Z bosons
particles that mediate the weak nuclear force

9780170459174

Revision summary

Use the following summary of syllabus dot points and key knowledge within Unit 4 Topic 3 to ensure that you have thoroughly reviewed the content. Provide a brief definition or comment for each item to demonstrate your understanding or code them using the traffic light system – Green (all good); Amber (needs some review); Red (priority area to review).

The cognitive verbs have been identified in bold.

The Standard Model	
• **define** the concept of an elementary particle and antiparticle	
• **recall** the six types of quarks	
• **define** the terms *baryon* and *meson*	
• **recall** the six types of leptons	
• **recall** the four gauge bosons	
• **describe** the strong nuclear, weak nuclear and electromagnetic forces in terms of the gauge bosons	
• **contrast** the fundamental forces experienced by quarks and leptons	
Particle interactions	
• **define** the concept of lepton number and baryon number	
• **recall** the conservation of lepton number and baryon number in particle interaction	
• **explain** the following interactions of particles using Feynman diagrams – electron and electron – electron and positron – a neutron decaying into a proton	
• **describe** the significance of symmetry in particle interactions	

© State of Queensland (QCAA) Physics General Senior Syllabus 2019 CC BY 4.0 https://creativecommons.org/licenses/by/4.0/

Exam practice

Multiple-choice questions

Each multiple-choice question is worth 1 mark.

Solutions start on page 140.

Question 1

The gauge bosons in the particle model include:

A hadrons, leptons and quarks.

B photons, neutrinos and leptons.

C protons, W boson, Z boson, antineutrinos.

D photons, W^+ boson, W^- boson, Z boson and gluons.

Question 2

Identify which of the following best defines the term 'lepton'.

A the number of quarks less the number of antiquarks in a composite particle

B an elementary particle of half-integer spin that undergoes strong interactions

C a subatomic particle which is intermediate in mass between an electron and a proton

D a family of elementary particles that includes electrons, taus, muons, their neutrinos and all of their antiparticles

Question 3

Identify which of the following Feynman diagrams displays an electron and positron annihilation.

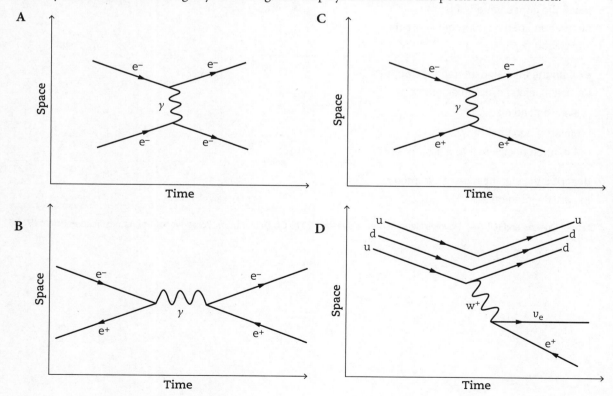

Question 4

Symmetry allows possible particle interactions to be determined. Identify which of the following statements includes examples of symmetry in particle interactions.

A baryon conservation and lepton conservation

B translation, reflection and rotation in time or space

C time-reversal, charge-reversal and crossing symmetry

D lepton conservation, charge-reversal and crossing symmetry

Question 5

Identify which of the following best defines the meson.

A a subatomic particle that includes a combination of up and down quarks only

B a subatomic particle of neutral charge and that includes the neutron

C a subatomic particle of less mass than an electron

D a family of heavy subatomic particles that contain one quark and one antiquark

Question 6

Identify which of the following Feynman diagrams displays an electron and electron interaction.

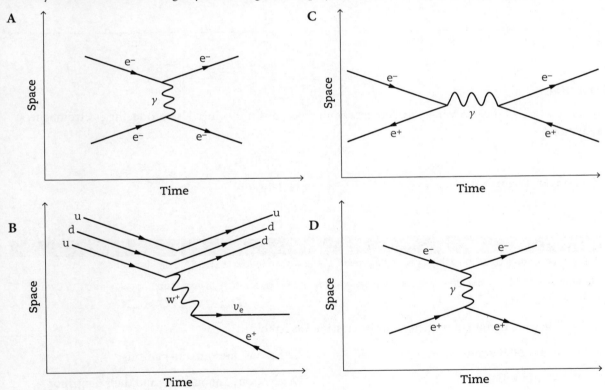

Question 7

Identify which of the following Feynman diagrams displays a beta decay.

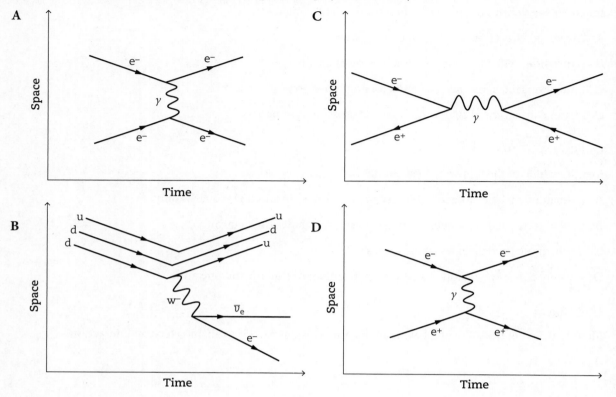

Question 8

Identify which of the following particles may annihilate a positron whilst also producing electromagnetic radiation?

A electron

B ultraviolet light

C positron

D neutron

Question 9

> **Hint**
>
> When answering a question asking you to *determine*, remember to follow the QCAA definition.
>
> **determine:** establish, conclude or ascertain after consideration, observation, investigation or calculation; decide or come to a resolution

Determine which group of particles may be classified as leptons.

A fermions and baryons

B quarks and antiquarks

C bosons, mesons and hadrons

D electrons, muons, taus and their neutrinos

Question 10

> **Hint**
>
> When answering a question asking you to *recall*, remember to follow the QCAA definition.
>
> **recall:** remember; present remembered ideas, facts or experiences; bring something back into thought, attention or into one's mind

Recall which particle is made up of two 'up' quarks and one 'down' quark.

A atom

B electron

C neutron

D proton

Short response questions

Question 11 (3 marks)

Describe the significance of symmetry in particle interactions. Include an example with an accompanying explanation.

> **Hint**
> When answering a question asking you to *describe*, remember to follow the QCAA definition.
> **describe:** give an account (written or spoken) of a situation, event, pattern or process, or of the characteristics or features of something

Question 12 (2 marks)

Define the concept of a baryon number and provide an example of a common baryon and its component quarks.

> **Hint**
> When answering a question asking you to *define*, remember to follow the QCAA definition.
> **define:** give the meaning of a word, phrase, concept or physical quantity; state meaning and identify or describe qualities

Question 13 (5 marks)

a Sketch the Feynman diagram of an electron–positron annihilation. Include and label all
components.   3 marks

> **Hint**
> When answering a question asking you to *sketch*, remember to follow the QCAA definition.
> **sketch:** execute a drawing or painting in simple form, giving essential features but not necessarily with detail or accuracy; in mathematics, represent by means of a diagram or graph; the sketch should give a general idea of the required shape or relationship and should include features

b Draw a possible crossing symmetry interaction of the Feynman diagram below.   2 marks

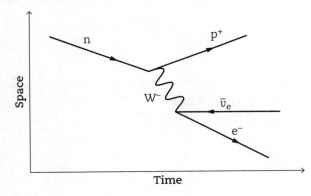

Question 14 (4 marks)

a   The concept of lepton number was introduced to assist in explaining the Cowan-Reines neutrino experiment, which investigated the two possible reactions below.

   i $\bar{v}_e + n \rightarrow p + e^-$

   ii $\bar{v}_e + p \rightarrow n + e^+$

Predict which of the reactions (i) or (ii) violated the conservation of lepton number.
Justify your response.   2 marks

> **Hint**
> When answering a question asking you to *predict*, remember to follow the QCAA definition.
> **predict:** give an expected result of an upcoming action or event; suggest what may happen based on available information

b   Baryon number is defined by $B = \dfrac{1}{3}(n_q - n_{\bar{q}})$, where n_q is the number of quarks and $n_{\bar{q}}$ is the number of antiquarks. Use this definition to determine the baryon number of the following particles.   2 marks

Particle	Quark composition	Baryon number
quark	q	
antiquark	$\bar{q}$	
proton	uud	
neutron	udd	
antiproton	$\bar{u}\bar{u}\bar{d}$	
antineutron	$\bar{u}\bar{d}\bar{d}$	

Question 15 (2 marks)

Recall all six (6) types of quarks and state their corresponding anti-quarks.

Question 16 (4 marks)

Identify which of the following particle interactions are possible through applying the laws of conservation of lepton number and conservation of baryon number.

a $p + n \rightarrow p + p + n + \bar{p}$ 2 marks

b $\bar{\mu} \rightarrow e^- + \bar{v}_e + v_\mu$ 1 mark

c $p + n \rightarrow p + p + \bar{p}$ 1 mark

Question 17 (6 marks)

Sketch Feynman diagrams to display:

a a neutron decaying into a proton.   3 marks

b an electron–positron annihilation   1 mark

c electron–electron scattering.   2 marks

Question 18 (4 marks)

Draw a table to represent the Standard Model, including the quarks, leptons and gauge bosons.

Question 19 (4 marks)

The decay of a neutron to a proton may be shown using a Feynman diagram. Use the axes below to show this interaction.

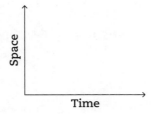

Question 20 (4 marks)

a Recall the exchange particles that mediate each of the forces below.   3 marks

Force	Gauge bosons
Electromagnetic	
Strong nuclear	
Weak nuclear	

b Explain why particles such as the quarks are termed fundamental particles.   1 mark

> **Hint**
>
> When answering a question asking you to *explain*, remember to follow the QCAA definition.
>
> **explain:** make an idea or situation plain or clear by describing it in more detail or revealing relevant facts; give an account; provide additional information

SOLUTIONS
UNIT 3: GRAVITY AND ELECTROMAGNETISM
CHAPTER 1 TOPIC 1: GRAVITY AND MOTION

Multiple-choice questions

Question 1 B

Normal force is defined as the force perpendicular to the surface that the object is on. B is correct.

Question 2 D

Using the formulas $u_x = u \cos \theta$ and $u_y = u \sin \theta$, the correct answer is D.

Question 3 D

Since centripetal force is proportional to velocity squared, the graph should be an increasing exponential. D is the correct answer.

Question 4 A

Using motion equations, the time of flight of the object is 1.56 s. From this, the initial horizontal velocity is calculated to be A.

Question 5 C

Using the formula $v = \dfrac{2\pi r}{T}$ and correctly calculating T to be 1.8 s, C is the correct answer.

Question 6 A

According to Kepler's second law of planetary motion, the closer the planet is to a star, the faster it will travel. Therefore, when the planet is furthest from the star, it will travel at its slowest speed. The only answer that provides a velocity slower than the average velocity is answer A.

Question 7 B

Since the object in Figure 1.25 is experiencing circular motion, its acceleration will be centre-seeking. The only answer that depicts a centre-seeking acceleration is answer B.

Question 8 D

The diagram shown demonstrates that at the beginning of the projectile's motion, the vertical velocity is positive (i.e. the object is travelling upwards). The correct answer is D as the remaining answers describe a negative or 0 vertical velocity at the beginning of their motion.

Question 9 D

The mass of the person is found to be 56.1 kg using the formula $F_w = mg_{Earth}$. The g on the Moon is found to be 1.62 m s^{-2} using the formula $g = \dfrac{Gm}{r^2}$. As a result, the weight of the person on the surface of the Moon is calculated using $F_w = mg_{Moon}$ and is answer D.

Question 10 B

Using the relationship $\dfrac{T_{Mercury}^2}{R_{Mercury}^3} = \dfrac{T_{Earth}^2}{R_{Earth}^3}$ (since the two planets orbit the same body), the period of Mercury is calculated to be answer B.

Short response questions

Question 11 (2 marks)

$$\frac{T^2}{r^3} = \frac{4\pi^2}{GM}$$

$$M = \frac{4\pi^2 r^3}{GT^2} = \frac{4\pi^2 (150 \times 10^9)^3}{6.674 \times 10^{-11} \times (365.25 \times 24 \times 60 \times 60)^2} = 2.00 \times 10^{30} \text{ kg}$$

Mark breakdown:
- 1 mark for manipulation of equation (or numerical values) to determine the mass of the Sun
- 1 mark for substitution of relevant value with units

Question 12 (4 marks)

a

$F_w = mg = 0.0755 \times 9.81 = 0.741 \text{ N}$

$F_N = F_w \text{ perpendicular} = mg \cos \theta = 0.741 \cos 40° = 0.567 \text{ N}$

$F_w \text{ parallel} = mg \sin \theta = 0.0755 \times 9.81 \times \sin 40° = 0.476 \text{ N}$

$F_t = m_{hanger} \times g = 0.488 \times 9.81 = 4.787 \text{ N}$

Mark breakdown:
- 1 mark for correct forces, without additions. Components of weight may be shown
- 1 mark for correct calculations for F_w, (or F_w perpendicular and F_w parallel) and F_t

b

Angle, θ (°)	Hanging mass (kg)	$\sin \theta$	Force weight parallel $m \times g \times \sin \theta$
10.0	0.131	0.174	0.128
20.0	0.263	0.342	0.253
30.0	0.379	0.500	0.370
40.0	0.488	0.643	0.476
50.0	0.588	0.766	0.567

The values for the parallel component of the weight and the weight of the hanging mass are very similar, so the data support their hypothesis that the weight of the hanging mass would be equal to the parallel component of the trolley cart's weight down the incline for each angle.

Mark breakdown:
- 1 mark for correct use of trigonometry and correct values (in table or working)
- 1 mark for reasoned analysis and statement regarding hypothesis

Question 13 (4 marks)

a

$$F = \frac{GMm}{r^2} = \frac{6.67 \times 10^{-11} \times 5.98 \times 10^{24} \times 2000}{(5.50 \times 10^6)^2}$$
$$= 26371\,\text{N}$$

Mark breakdown:
- 1 mark for correct substitution of values
- 1 mark for correct final force value

b

$$\frac{T^2}{r^3} = \frac{4\pi^2}{GM}$$
$$T^2 = \frac{4\pi^2 r^3}{GM} = \frac{4\pi^2 (6.37 \times 10^6 + 8.00 \times 10^6)^3}{6.67 \times 10^{-11} \times 5.98 \times 10^{24}}$$
$$T = 17138\,\text{s} = 4.76\ \text{hours}$$

Mark breakdown:
- 1 mark for correct algebraic or numerical manipulation
- 1 mark for substitution and unit conversions to determine final value

Question 14 (2 marks)

$$|v|^2 = |v_x|^2 + |v_y|^2$$
$$v_y = \sqrt{v^2 - v_x^2}$$
$$= \sqrt{150^2 - 62^2}$$
$$= 136.6\ \text{m s}^{-1}$$

$$\theta = \tan^{-1}\left(\frac{136.6}{62}\right) = 65.6° \text{ below the horizontal}$$

Mark breakdown:
- 1 mark for v_y calculated result
- 1 mark for θ determined below horizontal

Question 15 (4 marks)

On the assumption that friction is negligible, the accelerating force is due to the components of the force weight and force tension acting along the plane.

$$F_{\text{net}} = T \cos \theta - mg \sin \alpha$$
$$T = \frac{ma + mg \sin \alpha}{\cos \theta} = \frac{80 \times 1.0 + 80 \times 9.81 \times \sin 10°}{\cos 15°} = 224\ \text{N}$$

Mark breakdown:
- 1 mark for component of mg
- 1 mark for component of T
- 1 mark for algebraic or other rearranging
- 1 mark for correct answer

Question 16 (5 marks)

x-components

$F_{1x} = 0.00 \, \text{N}$

$F_{2x} = 30.00 \sin 20° = 10.26 \, \text{N}$

$F_{3x} = -15.0 \sin 20° = -5.13 \, \text{N}$

Thus, $\Sigma F_x = 5.13 \, \text{N}$ 1 mark

y-components

$F_{1y} = -15.00 \, \text{N}$

$F_{2y} = 30.00 \cos 20° = 28.19 \, \text{N}$

$F_{3y} = 15.0 \cos 20° = 14.10 \, \text{N}$

Thus, $\Sigma F_y = 27.29 \, \text{N}$ 1 mark

$|F_{net}| = \sqrt{5.13^2 + 27.29^2} = 27.77 \, \text{N}$ 1 mark

$\theta = \tan^{-1}\left(\dfrac{27.29}{5.13}\right) = 79.4°$ above the horizontal (or a true bearing of $010.6°$)   1 mark

$a = \dfrac{F_{net}}{m} = \left(\dfrac{27.77}{25}\right) = 1.11 \, \text{m s}^{-2}$ 1 mark

Question 17 (8 marks)

a $u_x = u_y = |u| \cos 45°$ (or $|U| \sin 45°$)

$s_y = u_y t + \dfrac{1}{2}a_y t^2$

$0 = u_x t - 4.9t^2$ 1 mark

Since $s_x = u_x t$

$t = \dfrac{s_x}{u_x}$ 1 mark

$0 = u_x\left(\dfrac{s_x}{u_x}\right) - 4.9t^2$

$4.9t^2 = 96.90$

$t = \sqrt{\dfrac{96.90}{4.9}} = 4.447 \, \text{s}$ 1 mark

b $u_y = u_x = \dfrac{s_x}{t} = \dfrac{96.90}{4.447} = 21.27 \, \text{m s}^{-1}$ 1 mark

$s_y = u_y t - 4.9t^2 = 21.79 \times \dfrac{4.447}{2} - 4.9 \times \left(\dfrac{4.447}{2}\right)^2$ (substitute t = half the time of flight)   1 mark

$s_y = 24.23 \, \text{m}$ 1 mark

c $|u| = \sqrt{u_x^2 + u_y^2} = \sqrt{21.79^2 + 21.79^2}$

$|u| = 30.82 \, \text{m s}^{-1}$ 1 mark

d using symmetry, $|v| = 30.82 \, \text{m s}^{-1}$ at 45° below the horizontal   1 mark

Question 18 (6 marks)

a $v = c = 3 \times 10^8$ m s^{-1}

$s = vt = 3 \times 10^8 \times 0.065$

$s = 1.95 \times 10^7$ m   1 mark

b $m_s = 200$ kg

$m_E = 5.98 \times 10^{24}$ kg

$r =$ altitude + radius of Earth

$= 0.065 \times 3.0 \times 10^8 + 6.37 \times 10^6$

$= 1.95 \times 10^7 + 6.37 \times 10^6$ m

$= 2.587 \times 10^7$ m   1 mark

$F_g = \dfrac{GMm}{r^2} = \dfrac{6.67 \times 10^{-11} \times 5.98 \times 10^{24} \times 200}{(2.587 \times 10^7)^2}$ 1 mark

$F_g = 119.2$N   1 mark

c A satellite performs uniform circular motion, thus the radius of the circle doesn't change nor does its speed.   1 mark

Although the magnitude of the speed remains the same the direction is continuously changing as the satellite is being constantly accelerated inwards, so it can maintain its circular path.   1 mark

Question 19 (6 marks)

a 1 mark for each correct law stated.

The first law states that all planets move about the Sun in elliptical orbits, having the Sun as one of the foci.

The second law states that a radius vector joining any planet to the Sun sweeps out equal areas in equal lengths of time.

The third law states that the squares of the sidereal periods of the planets are directly proportional to the cubes of their mean distance from the Sun.

b

Average period of orbit, $T = \dfrac{96.2 + 95.4 + 91.0}{3} = 94.2$ min

Mean altitude $= \dfrac{219 + 216 + 210}{3} = 215$ km   2 marks

c Number of orbits $= \dfrac{\text{total time}}{T}$

$= \dfrac{90 \text{ days} \times 24 \text{ hours per day} \times 60 \text{ min per hour}}{94.2 \text{ min per orbit}}$

$= 1380$ orbits   1 mark

Question 20 (3 marks)

$F = \dfrac{GMm}{r^2}$ 1 mark

$F = \dfrac{6.67 \times 10^{-11} \times 9.11 \times 10^{-31} \times 2 \times 1.67 \times 10^{-27}}{(1.5 \times 10^{-9})^2}$ 1 mark

$F = 9.02 \times 10^{-50}$ N   1 mark

UNIT 3: GRAVITY AND ELECTROMAGNETISM

CHAPTER 2 TOPIC 2: ELECTROMAGNETISM

Multiple-choice questions

Question 1 C

The current in Coil A travels in a clockwise direction and generates a magnetic field into the page. Coil B lies in this field with the field acting into the page, hence it experiences a current in the same direction. Therefore, the current in Coil B will travel in a clockwise direction.

Question 2 C

In Coulomb's law, force is inversely proportional to radius squared. Thus, if the distance between the two charges is tripled, the magnitude of the force on each sphere will be reduced by a factor of 9. The answer is C.

Question 3 C

Electromagnetic radiation is emitted in the form of waves or photons from an electromagnetic field. It does not require a supporting medium and can travel through space. The best description is answer C.

Question 4 C

For the conducting wire to be suspended, the gravitational force must be balanced by the magnetic force (i.e. the net force must equal 0). If the gravitational force always acts downwards, then the magnetic force must act upwards. Using the right-hand rule, it can be determined that the magnetic field travels out of the page and that the answer is C.

Question 5 B

Using the formula $B = \mu_0 nI$, and ensuring that n = turns/metre, the answer is calculated to be B.

Question 6 D

Electromagnetic induction involves the production of an electric field by a changing magnetic field. An electric field means there is an electromotive force (emf) acting between points in that region. If this emf acts upon free charge carriers, an induced current is generated, but only if the coil forms a complete loop. Thus, the correct answer is D.

Question 7 B

Using the formula $\lambda = c/f$ and correctly converting units of wavelength, the correct answer is B.

Question 8 C

When a magnet is pushed into a coil, a current will be induced in the coil that opposes this motion, according to Lenz's law. Therefore, the correct answer is C.

Question 9 B

Applying the right-hand rule, with the thumb in the direction of the conventional current and the fingers in the direction of the external magnetic field the palm faces into the page.

Question 10 B

Since the charges are attractive (i.e. they have opposite polarities), the force between the two charges will be negative. Using the formula $F = \dfrac{1}{4\pi\varepsilon_0}\dfrac{Qq}{r^2}$, and correctly converting the charges to coulombs, answer B is obtained.

Short response questions

Question 11 (4 marks)

a The alternating current in the primary coil generates a changing magnetic field. [1 mark]

Placing a secondary conducting coil in this changing field results in the flux through the secondary coil changing over time [1 mark], thereby inducing a current in the secondary coil (to oppose the change in flux). [1 mark] [3 marks in total]

b

$$\frac{V_P}{V_S} = \frac{n_P}{n_S}$$

$$n_S = n_P \frac{V_S}{V_P} = 80 \times \frac{60}{240} = 20 \text{ turns}$$

The secondary coil has 20 windings.

Rearrangement and substitution   1 mark

Question 12 (4 marks)

a As the change in flux is downwards, the induced field would be upwards in opposition.   1 mark

Hence current would flow clockwise from X to Y, using the right-hand rule.
(counter clockwise when viewed from above).   1 mark

b magnetic flux $= \Delta\Phi = \Delta BA \cos\theta = -0.2 \times (0.08 \times 0.24) \cos 0° = 3.84 \text{ mWb}$

> **Mark breakdown:**
> - 1 mark for correct substitution and rearrangement
> - 1 mark for correct final result including unit

Question 13 (4 marks)

a Using the right-hand rule, if the particle were positively charged, an inward centripetal force would be generated if the magnetic field is into the page. As the charge is negative, the direction of the magnetic field is opposite and is therefore out of the page.   2 marks

b Equating centripetal force and magnetic force:

$$\frac{mv^2}{r} = qvB \sin\theta$$

Hence $r = \dfrac{mv}{qB \sin\theta}$

  i $r \propto v$, so radius would increase by a factor of 3.   1 mark

  ii $r \propto \dfrac{1}{q}$, so radius would halve if the magnitude of the charge were doubled.   1 mark

Question 14 (4 marks)

a

$$E = \frac{1}{4\pi\varepsilon_0} \frac{q}{r^2}$$

at R. At S, the distance has quadrupled, so the electric field strength has reduced by a factor of 4^2 or 16. Hence, at S, $E = 1.125 \text{ N C}^{-1}$.

Mark breakdown:
- 1 mark for calculation of E at S
- 1 mark for accompanying working

b

Since $F = \dfrac{kqQ}{d^2} = qE$, then q may be determined using either point and $k = 9 \times 10^9$.

Hence $q = E \times r^2 \times \dfrac{1}{k}$

And $q = 18 \times 1^2 \times \dfrac{1}{9 \times 10^9}$ at R

So $q = 2.0 \times 10^{-9}$ C

$E = \dfrac{kq}{d^2}$

$10 = \dfrac{9 \times 10^9 \times 2.0 \times 10^{-9}}{d^2}$

$d = 1.34$ m

Mark breakdown:
- 1 mark for calculation of the magnitude of charge q
- 1 mark for calculation of radius

Question 15 (3 marks)

a $F = qvB \sin \theta = 1.6 \times 10^{-19} \times 2.5 \times 10^5 \times 3.0 \times 10^{-2} \times 1 = 1.2 \times 10^{-15}$ N

Mark breakdown:
- 1 mark for correct calculation of value

b   1 mark

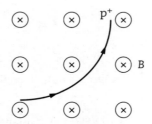

FIGURE 7.1 The direction of a proton's circular path is consistent with the right-hand rule for a positive charge in an external magnetic field.

c

Since $F_c = F_B$

So, $a_c = \dfrac{F_B}{m} = \dfrac{1.2 \times 10^{-15}}{1.67 \times 10^{-27}} = 7.19 \times 10^{11}$ m s^{-2} 1 mark

Question 16 (2 marks)

$B = \mu_0 nI = 1.26 \times 10^{-6} \times \dfrac{600}{0.4} \times 500 \times 10^{-3} = 9.45 \times 10^{-4}$ T

Mark breakdown:
- 1 mark for correct use of turns per metre in substitution
- 1 mark for correct calculation of result

Question 17 (2 marks)

a Doubling the magnetic field also doubles the magnetic flux, and hence the rate of change. The waveform will thus double in amplitude.　　　1 mark

b Doubling the rate of rotation doubles the rate of change as it cycles twice as often. The amplitude thus doubles (and the period halves).　　　1 mark

Question 18 (4 marks)

a $L = 0.40\,\text{m}$, $B = 0.6\,\text{T}$, $I = 1.5\,\text{A}$, $F = 0.18\,\text{N}$

$F = BIL \sin \theta$

$$\theta = \sin^{-1}\left(\frac{F}{BIL}\right)$$　　　1 mark

$$\theta = \sin^{-1}\left(\frac{0.18}{0.6 \times 1.5 \times 0.4}\right)$$　　　1 mark

$\theta = 30.0°$　　　1 mark

b Using the right-hand rule with the magnetic field B facing up and conventional current, I, running towards the right, the magnetic force, F, is out of the page.　　　1 mark

Question 19 (3 marks)

$\sqrt{\text{AB}} = \sqrt{1^2 + 1^2} = \sqrt{2}$ m, so

$$|F_{\text{AB}}| = \frac{1}{4\pi\varepsilon_0}\frac{q_1 q_2}{\sqrt{\text{AB}}^2} = 0.0405\,\text{N}$$

$F_{\text{AB}_x} = 0.0405$ N, $F_{\text{AB}_y} = 0$ N　　　1 mark

$$|F_{\text{CB}}| = \frac{1}{4\pi\varepsilon_0}\frac{q_1 q_2}{\sqrt{\text{BC}}^2} = -0.081\,\text{N}$$

$F_{\text{CB}_x} = -0.081 \cos 45° = -0.0573$ N　　　1 mark

$F_{net_x} = 0.0405 - 0.0573 = -0.0168$ N

$F_{net_y} = -0.081 \sin 45° = -0.0573$ N　　　1 mark

Question 20 (5 marks)

a Gradient $= 42.3\,\text{g/A}$ or $4.23 \times 10^{-2}\,\text{kg/A}$. Represents how strongly the force (or mass) depends on current through the wire.　　　1 mark

y-intercept is zero current, i.e. no force due to magnetic field, hence this is the mass of the coil.

$m = 108\,\text{g}$, weight $= mg = 1.1\,\text{N}$

[Correct mass and weight 1 mark]

b $F = BIL \sin \theta$ $(\theta = 90°)$

$mg = BIL$

$$m = \frac{BL}{g} \times 1$$

Thus, gradient $= \dfrac{BL}{9}$

$$B = \frac{\text{gradient} \times g}{L} = \frac{4.23 \times 10^{-2} \times 9.8}{0.04 \times 25} = 0.415\,\text{T}$$　　　1 mark

c $F = BIL \sin \theta = 1.1\,\text{N}$

$$I = \frac{F}{BL \sin \theta} = \frac{1.1}{0.415 \times 1 \times 1} = 2.7\,\text{A}$$　　　1 mark

Current is directed out of the page along the bottom of the coil.　　　1 mark

UNIT 3: GRAVITY AND ELECTROMAGNETISM

CHAPTER 3 UNIT 3 DATA TEST

Data set 1

Question 1

$$\frac{8.35 + 8.52 + 8.31}{3} = 8.39$$

Average force = 8.39 μN (2 d.p.)

> **Mark breakdown:**
> - 1 mark for the correct solution

Question 2

$$\pm \frac{x_{max} - x_{min}}{2}$$

$$= \pm \frac{8.52 - 8.31}{2}$$

$$= \pm 0.105 \text{ μN}$$

Absolute uncertainty = ±0.11 μN (2 d.p.)

> **Mark breakdown:**
> - 1 mark for correct solution
> - 1 mark for working

Question 3

As the distance increases by a factor of 2, the average force decreases by a factor of 4; for example, when the distance increases from 0.020 m to 0.040 m, the force changed from 8.39 μN to 2.13 μN.

This is a decrease by a factor of 4 approximately.

There is an inverse square relationship between force and distance.

OR

$$F \propto \frac{1}{d^2}$$

(or words to that effect, including supporting example)

> **Mark breakdown:**
> - 1 mark for identification of the relationship
> - 1 mark for an appropriate justification using an example.

Question 4

As distance doubles from 0.040 m to 0.080 m, it is expected that the force would approximately quarter in value.

$$F = \frac{1}{4} \times 2.13 \ \mu N = 0.53 \ \mu N$$

OR similar use of values from table

Mark breakdown:
- 1 mark for identifying the relationship
- 1 mark for reasoning/working.
- 1 mark for solution

Data set 2

Question 1

$u_h = u \times \cos \theta$

$u_h = 10.0 \times \cos 40°$

$u_h = 7.66 \ m\,s^{-1}$

Horizontal component of velocity = $7.66 \ m\,s^{-1}$ (2 d.p.)

Mark breakdown:
- 1 mark for working. No mark for formula without substitution
- 1 mark for horizontal component of velocity

Question 2

Maximum height = 0.50 m (read from graph)

Mark breakdown:
- 1 mark for identifying the value. Accept a tolerance of ± 0.1 m

Question 3:

When the initial velocity doubles from 2 m s^{-1} to 4 m s^{-1}, the maximum height increases by a factor of 3.3 (or approximately 4 or 2^2) from 0.15 m to 0.50 m.

When the initial velocity doubles from 4 m s^{-1} to 8 m s^{-1}, the maximum height increases by a factor of 3.5 (approximately 4) from 0.50 m to 2.00 m.

Hence, it can be determined that the maximum height is proportional to the initial velocity squared.

Mark breakdown:
- 1 mark for identifying the relationship
- 1 mark for use of data to justify. Any pair of values acceptable

Question 4

As the distance increases by a factor of 2, the maximum height reached increases by a factor of 4; for example, when the velocity increases from $4\,\text{m s}^{-1}$ to $8\,\text{m s}^{-1}$, the maximum height reached increased from $0.5\,\text{m}$ to $2.0\,\text{m}$.

This relationship is a squared relationship between maximum height reached and initial velocity.

Further, as the distance increases by a factor of 2, the maximum height reached increases by a factor of 4; hence, when the velocity increases from $6\,\text{m s}^{-1}$ to $12\,\text{m s}^{-1}$, the maximum height reached is predicted to increase from $1.0\,\text{m}$ to $4.0\,\text{m}$.

The predicted maximum height for a velocity of $12\ \text{m s}^{-1}$ is $4.0\,\text{m}$.

Mark breakdown:
- 1 mark for correct value of $4.0\,\text{m}$ ($\pm\,0.2\,\text{m}$)
- 1 mark for an appropriate justification using an example.

Data set 3

Question 1:

$$\text{Uncertainty in gradient} = \frac{(\text{maximum gradient} - \text{minimum gradient})}{2}$$

$$\frac{(0.3117 - 0.2498)}{2} = 0.0310$$

$$\text{Percentage uncertainty in gradient} = \frac{\text{Uncertainty in gradient}}{\text{Average measured value of gradient}} \times 100\%$$

$$= \frac{0.0310}{0.2801} \times 100\%$$

$$= 11.0\% \text{ (1 d.p.)}$$

Mark breakdown:
- 1 mark for the correct uncertainty in gradient
- 1 mark for the correct percentage uncertainty in gradient determination

Question 2

The relationships is an inverse square relationship between force and distance, or $F \propto \dfrac{1}{d^2}$.

This is evident from the graph as well as the table of data.

The question requires reference to values from the table of data.

As the distance increases by a factor of 2, the average force decreases by a factor of 4; for example, when the distance increases from $0.0050\,\text{m}$ to $0.0100\,\text{m}$, the force changed from $115.24\ \mu\text{N}$ to $28.23\ \mu\text{N}$.

This is a decrease by a factor of 4 approximately.

Mark breakdown:
- 1 mark for identification of the relationship
- 1 mark for an appropriate justification using an example with values from the table. Accept other, similar reasoning

Question 3

As distance doubles from 0.125 m to 0.250 m it is expected that the force would approximately quarter its value.

$$F = \frac{1}{4} \times 22.12 \ \mu N = 5.53 \ \mu N$$

OR similar use of values from table

Mark breakdown:
- 1 mark for identifying the relationship
- 1 mark for reasoning/working.
- 1 mark for solution

UNIT 4: REVOLUTIONS IN MODERN PHYSICS
CHAPTER 4 TOPIC 1: SPECIAL RELATIVITY

Multiple-choice questions

Question 1 D

High-speed muons formed in Earth's upper atmosphere should not last long enough to reach Earth's surface. However, muons are observed near Earth's surface. The moving muons have a longer lifetime than stationary muons. This cannot be explained by Newtonian physics.

Question 2 C

A reference frame where Newton's first law of inertia applies is valid.

Question 3 C

Length contracts for the observer travelling at high speeds on board the spaceship. Thus, the spacecraft is observed to be shorter than 190 m. Using the length contraction equation, answer C is obtained.

Question 4 D

Answer D identifies two postulates correctly: the relativity principle and the constancy of the speed of light principle.

Question 5 D

What is perceived in one reference frame is not necessarily the same as another, but both can still be correct, as long as special relativity principles are upheld. D is the correct answer.

Question 6 C

C is the correct answer as it identifies that the reference frame must not be accelerating, instead maintaining inertia.

Question 7 D

For Einstein's postulates of special relativity to be true, two events that are simultaneous for one observer are not necessarily simultaneous for another. This supports answer D.

Question 8 C

The correct rearrangement is C. Other solutions are incorrectly rearranged.

Question 9 B

The change in mass for the reaction is 0.22535 u or 3.742×10^{-25} kg. Using the mass–energy equivalence statement, the answer is B.

Question 10 D

Time dilates for the observer on the platform. As such, the time observed should be D using the time dilation equation.

Short response questions

Question 11 (3 marks)

a Newtonian: $p = mv = 9.109 \times 10^{-31} \times 0.992 \times 3 \times 10^8 = 2.71 \times 10^{-22}\,\text{kg m s}^{-1}$

Relativistic momentum $= \dfrac{mv}{\sqrt{1 - \dfrac{v^2}{c^2}}} = \dfrac{9.109 \times 10^{-31} \times 0.992 \times 3 \times 10^8}{\sqrt{1 - 0.992^2}} = 2.15 \times 10^{-21}\,\text{kg m s}^{-1}$

Mark breakdown:
- 1 mark for correct numerical value of Newtonian momentum
- 1 mark for correct numerical value of relativistic momentum

b Ratio $= \dfrac{2.15 \times 10^{-21}}{2.71 \times 10^{-22}} = 7.93$

The relativistic value is 7.93 times larger than the Newtonian value.

Mark breakdown:
- 1 mark for correct process

Question 12 (5 marks)

a **i** $V_b = V_{a\,\text{rel}\,b} + V_a$

$= 4 + 25$

$= 29\,\text{m s}^{-1}$

Mark breakdown:
- 1 mark for correct calculation of **i**

ii $V_{a\,\text{rel}\,b} = 4\,\text{m s}^{-1}$

Mark breakdown:
- 1 mark for correct identification of **ii**

b To an observer in the barn, the length of the pole of a person running towards the barn is contracted, hence it may fit in the barn (that is, both ends may fit simultaneously inside the barn).

In contrast, the person with the pole observes the barn length to be contracted in this dimension, hence the pole does not fit in the barn.

The resolution is that the simultaneous events for the observer in the barn and the person running with the pole occur at different times.

Mark breakdown:
- 2 marks for correct description of paradox
- 1 mark for provision of resolution

Question 13 (3 marks)

a For observers on Earth, $t = \dfrac{s}{v} = \dfrac{4.35\ \text{ly}}{0.75c} = 5.8$ years (or 183 000 000 s)

Note: $1\,\text{ly} = 9.46 \times 10^{15}\,\text{m}$

This is the dilated time, the time on the spacecraft is t_0.

$$t = \frac{t_0}{\sqrt{\left(1 - \frac{v^2}{c^2}\right)}}$$

$$t_0 = t\sqrt{\left(1 - \frac{v^2}{c^2}\right)} = 5.8 \times \sqrt{1 - 0.75^2} = 3.84 \text{ years (or } 121\,000\,000 \text{ s)}$$

Mark breakdown:
- 1 mark for correct calculated value for observers on Earth
- 1 mark for correct process to find t_0; alternatively use contracted length and $v = 0.75c$

b $L = L_0\sqrt{\left(1 - \frac{v^2}{c^2}\right)} = 4.25 \text{ ly} \times \sqrt{1 - 0.75^2} = 2.88 \text{ ly (or } 2.72 \times 10^{16}\,\text{m)}$

Mark breakdown:
- 1 mark for correct process to determine the contracted length, L

Question 14 (4 marks)

a The stationary observer measures the spacecraft to be shorter than its rest length (i.e. that measured by the traveller on the spacecraft). The observer measures a shorter length for objects that are moving relative to themselves. 1 mark

The traveller measures a shorter distance for any object at rest in the stationary observer's reference frame. 1 mark

b

$$L = L_0\sqrt{1 - \frac{v^2}{c^2}}$$

$$L_0 = \frac{L}{\sqrt{1 - \frac{v^2}{c^2}}} = \frac{1.10 \times 10^2}{\sqrt{1 - 0.8^2}} = 1.83 \times 10^2 \text{ m}$$

Mark breakdown:
- 1 mark for rearranging or equivalent
- 1 mark for correct calculation

Question 15 (2 marks)

The time measured in the rest frame for the events is the proper time (the shortest time interval).
Any reference frame moving with respect to this rest frame will observe an increased time for any event (the time between them) due to time dilation. 1 mark

Let $v = 0.9c$ and $t_0 = 100$ s.

$$t = \frac{t_0}{\sqrt{\left(1 - \frac{v^2}{c^2}\right)}}$$

$$t = \frac{100}{\sqrt{\left(1 - \frac{0.9^2\,c^2}{c^2}\right)}} = 229 \text{ s}$$

Time dilation calculation 1 mark

Question 16 (4 marks)

a
$$p = \frac{mv}{\sqrt{1 - \frac{v^2}{c^2}}} = \frac{2.48 \times 10^{-28} \times 0.994 \times 3 \times 10^8}{\sqrt{1 - 0.994^2}}$$ 1 mark

$$p = 6.76 \times 10^{-19} \text{ kg m s}^{-1}$$ 1 mark

b $p = mv = 2.48 \times 10^{-28} \times 0.994 \times 3 \times 10^8$ 1 mark

$$p = 7.40 \times 10^{-20} \text{ kg m s}^{-1}$$ 1 mark

Question 17 (4 marks)

$t_0 = 45.0 \text{ s}$ 1 mark

$$t = \frac{t_0}{\sqrt{1 - \frac{v^2}{c^2}}}$$ 1 mark

$$t = \frac{45.0}{\sqrt{1 - 0.85^2}}$$ 1 mark

$t = 85.4 \text{ s}$ 1 mark

Question 18 (5 marks)

a The speed of light is constant and is independent of the velocity of the source or the observer, hence both the passenger on the train and the stationary observer view the speed of light as c.   2 marks

b
$$L = L_0 \sqrt{1 - \frac{v^2}{c^2}}$$

$$L = 90.0 \times \sqrt{1 - \frac{v^2}{c^2}} = 90.0 \times \sqrt{1 - \frac{0.88^2 c^2}{c^2}} = 42.7 \text{ m}$$ 3 marks

Question 19 (3 marks)

Mu mesons (or muons) that are created in Earth's upper atmosphere have a very short resting lifetime (approximately 2.2 μs). They are so short-lived that they would not reach the surface of Earth without relativistic effects. As they travel at such high speeds, time dilation applies and their lifetimes are increased long enough to be able to be detected on the surface of Earth.   3 marks

> **Note:**
> Similar scenarios or phenomena may also be used for the answer.

Question 20 (4 marks)

a Postulate 1: the laws of physics are the same in all inertial reference frames.   1 mark

Postulate 2: the speed of light, c, is constant in all inertial reference frames and is independent of the velocity of the source or observer.   1 mark

b As an object approaches the speed of light its momentum increases due to relativistic effects. There is therefore a limit to the velocity, as it would require an infinite amount of energy to accelerate to this limit.   2 marks

UNIT 4: REVOLUTIONS IN MODERN PHYSICS

CHAPTER 5 TOPIC 2: QUANTUM THEORY

Multiple-choice questions

Question 1 D

Wein determined the law relating wavelength and temperature (in Kelvin) for black bodies.

Question 2 B

Using the formula $E = hf$ and values of $f = 5.45 \times 10^{14}$ Hz and $h = 6.626 \times 10^{-34}$ J s, answer B is obtained, once the value is correctly converted to electron volts.

Question 3 D

The charged electrons were continuously accelerating, hence radiating energy and thus losing kinetic and potential energy through the law of conservation of energy.

Question 4 D

The photoelectric effect is an experiment that involves light of a certain frequency striking a metal surface. Electrons are subsequently ejected from the metal.

Question 5 B

The constructive and destructive interference patterns support the wave nature of light.

Question 6 A

As the diagram demonstrates, a wavelength of 4.5×10^{-8} m lies within the ultraviolet region. The wavelength correctly converted into nanometres is 45 nm and the correct frequency of 6.67×10^{15} Hz is calculated using $f = c/\lambda$.

Question 7 C

Using the formulae $E_k = hf - w$ and $E_k = \frac{1}{2}mv^2$ and correctly calculating E_k to be 2.06×10^{-19} J, answer C is obtained.

Question 8 C

Question 9 B

Once the threshold frequency is met, each additional incident photon results in an additional electron being ejected from the surface, hence an increase in the photocurrent.

Question 10 C

The ratio of the maximum kinetic energy and frequency of incident light provides Planck's constant and is the same for all metals.

Short response questions

Question 11 (3 marks)

Transitions between energy levels in the Bohr model correspond to photon energies of the emitted light in the spectrum. For example, for the transition from level 5 to level 2, the energy difference is $(-0.87 - -5.43) \times 10^{-19}\,\text{J} = 4.56 \times 10^{-19}\,\text{J}$

Equating wavelength to the energy of a photon:

$$E = \frac{hc}{\lambda}$$

$$\lambda = \frac{hc}{E} = \frac{6.626 \times 10^{-34} \times 3 \times 10^{8}}{4.56 \times 10^{-19}} = 436\,\text{nm}$$

One of the visible lines of the hydrogen spectrum is 436 nm.

Mark breakdown:

- 2 marks for linking photons emitted due to transitions in electron energy levels
- 1 mark for sample wavelength calculation

Question 12 (3 marks)

a

$\Delta E = 2.9 \times 10^{-19}\,\text{J}$

This occurs when the electron moves between $n = 1$ and $n = 2$.

$\Delta E = -5.97 \times 10^{-19}\,\text{J} - -8.87 \times 10^{-19}\,\text{J} = 2.9 \times 10^{-19}\,\text{J}$

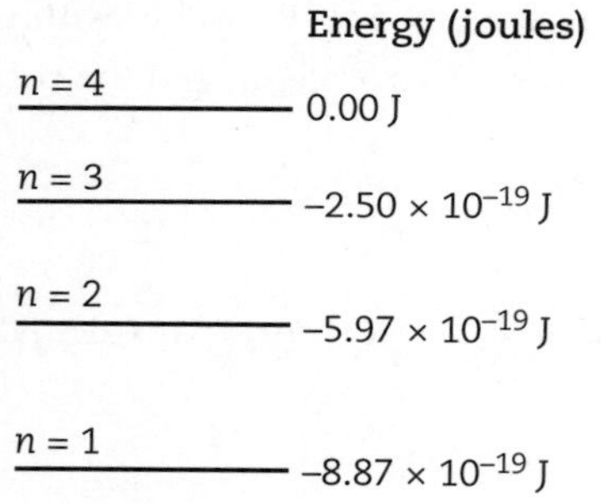

	Energy (joules)
$n = 4$	0.00 J
$n = 3$	-2.50×10^{-19} J
$n = 2$	-5.97×10^{-19} J
$n = 1$	-8.87×10^{-19} J

FIGURE 7.3 Energy level diagram for a mercury atom

Mark breakdown:

- 1 mark for an appropriate line
- 1 mark for supporting calculation

b Maximum frequency corresponds to maximum energy ($E = hf$), thus maximum frequency occurs between $n = 4$ and $n = 1$.

$\Delta E = -8.87 \times 10^{-19}\,\text{J}$

$$f = \frac{E}{h} = \frac{8.87 \times 10^{-19}}{6.626 \times 10^{-34}} = 1.34 \times 10^{15}\,\text{Hz}$$

Mark breakdown:

- 1 mark for recognition of greatest energy difference and calculation frequency

Question 13 (3 marks)

Light can act as either a particle or a wave, depending on how it is measured, hence the duality.   1 mark

The ejection of a photoelectron by light of a sufficiently high frequency in the photoelectric effect demonstrates the particle nature of light.   1 mark

Young's double-slit experiment, which demonstrates interference patterns, supports the wave behaviour of light.   1 mark

Question 14 (3 marks)

a $E = hf = \dfrac{hc}{\lambda}$ 1 mark

$E = \dfrac{6.626 \times 10^{-34} \times 3 \times 10^{8}}{6.50 \times 10^{-7}}$

$E = 3.06 \times 10^{-19}\,\text{J}$

$E = 1.91\,\text{eV}$ 1 mark

b $\lambda = \dfrac{hc}{E} = \dfrac{6.626 \times 10^{-34} \times 3 \times 10^{8}}{2.9 \times 10^{3} \times 1.6 \times 10^{-19}}$

$\lambda = 4.28 \times 10^{-10}$ m or 0.428 nm   1 mark

Question 15 (4 marks)

$E_{2\,\text{to}\,1} = 4.86\,\text{eV}$ 1 mark

$E_{2\,\text{to}\,1} = 7.776 \times 10^{-19}\,\text{J}$ 1 mark

$F = \dfrac{E}{h} = \dfrac{7.776 \times 10^{-19}}{6.626 \times 10^{-34}} = 1.17 \times 10^{15}\,\text{Hz}$ 1 mark

This is slightly higher than visible light, just inside the ultraviolet light region.   1 mark

Question 16 (2 marks)

There are two key elements to this explanation.

Black bodies emit a wide spectrum of electromagnetic thermal radiation at any temperature above 0 K.
    1 mark

The quantisation of light is required to avoid the ultraviolet catastrophe. Given that objects may only emit discrete photon energies, then the probability of a higher energy (UV) light emission is lowered, avoiding the ultraviolet catastrophe.   1 mark

Question 17 (2 marks)

Kinetic energy $= hf = \dfrac{1}{2}mv^{2}$ 1 mark

$M = \dfrac{2hf}{v^{2}} = \dfrac{2 \times 6.626 \times 10^{-34} \times 3.5 \times 10^{14}}{(2.10 \times 10^{4})^{2}} = 1.05 \times 10^{-27}\,\text{kg}$ 1 mark

Question 18 (6 marks)

a $E = hf$, $f = \dfrac{c}{\lambda}$

$$E = 6.626 \times 10^{-34} \times \dfrac{3 \times 10^8}{250 \times 10^{-9}} = 7.95 \times 10^{-19} \text{ J}$$

1 mark

$$W = 3.70 \text{ eV} = 3.70 \times 1.6 \times 10^{-19} = 5.92 \times 10^{-19} \text{ J}$$

1 mark

Since $W < E_{\text{photon}}$, the photoelectric effect occurs.

1 mark

b $E_{k(\text{max})} = hf - W = 7.95 \times 10^{-19} - 5.92 \times 10^{-19}$

$E_{k(\text{max})} = 2.03 \times 10^{-19} \text{ J}$

1 mark

c $E_k = \dfrac{1}{2}mv^2$

$$v = \sqrt{\dfrac{2E_k}{m}}$$

1 mark

$$v = \sqrt{\dfrac{2 \times 2.03 \times 10^{-19}}{9.1 \times 10^{-31}}} = 6.68 \times 10^5 \text{ m s}^{-1}$$

1 mark

Question 19 (5 marks)

a 1 mark each for any three of the relevant features below.

- the intensity of light is proportional to number of photoelectrons
- no photoelectrons are released below a threshold frequency of light
- the kinetic energy of photoelectrons increases with frequency of light
- photoelectrons are released immediately when light strikes the metal
- or other suitable responses consistent with the photoelectric effect

b 1 mark each for justification of how any two features provide evidence of photon quantisation, such as:

- the photoelectric effect experiment demonstrates that the number of photoelectrons ejected from a metal plate is proportional to incident light intensity, whereas the kinetic energy of the ejected electrons is proportional to the incident light frequency. The fact that frequencies lower than a threshold value will not eject an electron suggests that light energy is quantised
- as the frequency increases, the kinetic energy of the ejected photoelectrons increases, suggesting that light energy is quantised
- the fact that a greater intensity of light causes more electrons to be ejected suggests that the greater the intensity (more photons) incident on the metal plate the more electrons are ejected, suggesting that light energy is quantised.

Question 20 (4 marks)

The Rutherford model's limitations included: **a)** that a charged electron orbiting a positive nucleus should radiate electromagnetic radiation continuously, and **b)** that it did not explain the emission spectra of gases such as hydrogen.

2 marks

Bohr's model addressed these limitations by determining that the electrons followed several rules, including:

1 Electrons orbit in stable states without emitting electromagnetic radiation.

2 Different states represent different radii about the nucleus and the energy of electrons corresponds to these states.

3 Any change in state or orbit of an electron accompanies a change in energy (absorbing or emitting electromagnetic radiation equivalent to $E = hf$).

4 An electron's angular momentum is quantised and able to take only integer multiples of Planck's constant, where $L = mvr = \dfrac{nh}{2\pi}$.

[2 marks; $\dfrac{1}{2}$ mark each]

UNIT 4: REVOLUTIONS IN MODERN PHYSICS

CHAPTER 6 TOPIC 3: THE STANDARD MODEL

Multiple choice questions

Question 1 D

Answer D describes only force carrier particles called gauge bosons. Alternative answers describe other fundamental particles such as quarks and leptons.

Question 2 D

A lepton is a type of elementary particle that exists freely and interacts via the weak nuclear force. There are six types of leptons that are listed in answer D, the correct answer.

Question 3 C

Answers B and C are the only diagrams that display an electron (e^-) and positron (e^+) interaction. In answer B, the horizontal gauge boson indicates a period of time when the electron and positron do not exist, i.e. they annihilate. Thus, C is the correct answer to this question.

Question 4 C

Answers A and D describe conservation laws. While answer B lists symmetry operations, answer C responds directly to the question by providing specific examples of symmetry in particle interactions.

Question 5 D

A meson is defined as a subatomic, hadronic particle composed of one quark and one antiquark, as described in answer D.

Question 6 D

D is the correct answer as it contains the only Feynman diagram of an electron (e^-) interacting with another electron (e^-).

Question 7 B

Beta decay is when an unstable neutron (n) decays into a proton (p^+) and releases an intermediate particle, a virtual W^- boson, which transforms into an electron (e^-) and electron antineutrino ($\overline{v}_e$). Answer B best describes this process.

Question 8 A

Annihilation occurs when a subatomic particle and respective antiparticle (e.g. positron and electron) interact. The virtual photons formed from the annihilation exist only for a fraction of a second.

Question 9 D

The six leptons in the standard model are an electron, an electron neutrino, a muon, a muon neutrino, a tau and a tau neutrino. Answer D is the only answer that contains only leptons.

Question 10 D

A particle composed of three quarks is called a baryon. Out of the answers listed, answer D is the only baryon: a proton is comprised of two 'up' quarks and one 'down' quark.

Short response questions

Question 11 (3 marks)

Symmetry allows the prediction of pairs of interactions by swapping all charges (charge-reversal symmetry), reversing time, or by replacing a particle by its antiparticle on the opposite side of the interaction (crossing symmetry).

An example of time-reversal symmetry is that of the decay of a neutron into a proton by emitting an electron and an antineutrino. This could also happen in reverse i.e. if an antineutrino, an electron and a proton combine, a neutron can be formed.

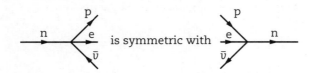

Mark breakdown:

- 1 mark for correct description of significance of symmetry in terms of particle interactions
- 1 mark for correct inclusion of at least one form of symmetry
- 1 mark for correct identification of example of symmetry provided

Question 12 (2 marks)

Baryon number represents the number of baryons in a system, minus the number of antibaryons.

A proton is a baryon that consists of two up quarks and one down quark.

Mark breakdown:

- 1 mark for correct definition
- 1 mark for correct example

Question 13 (5 marks)

a

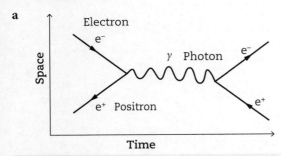

Mark breakdown:

- 1 mark for components and axes set out correctly
- 1 mark for correct directions or arrows and labels for electron and positron
- 1 mark for correct identification of photon

b

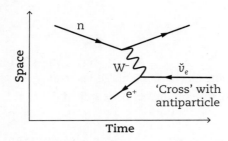

Mark breakdown:
- 1 mark for correct and plausible interaction identified
- 1 mark for components, axes and labels of particles set out correctly

Question 14 (4 marks)

a i Violates conservation of lepton number   1 mark

The left-hand side has an antineutrino ($L_e = -1$), while the right-hand side has an electron ($L_e = 1$)   1 mark

b

Particle	Quark composition	Baryon number
quark	q	$\dfrac{1}{3}$
antiquark	$\bar{q}$	$\dfrac{-1}{3}$
proton	uud	$\dfrac{1}{3} \times (3 - 0) = 1$
neutron	udd	$\dfrac{1}{3} \times (3 - 0) = 1$
antiproton	$\bar{u}\bar{u}\bar{d}$	$\dfrac{1}{3} \times (0 - 3) = -1$
antineutron	$\bar{u}\bar{d}\bar{d}$	$\dfrac{1}{3} \times (0 - 3) = -1$

Mark breakdown:
- 1 mark for correct baryon number values determined for the quark and antiquark
- 1 mark for correct baryon number values determined for the proton, neutron, antiproton and antineutron

Question 15 (2 marks)

up, down, charm, strange, top, bottom   1 mark

anti-up, anti-down, anti-charm, anti-strange, anti-top, anti-bottom   1 mark

Question 16 (4 marks)

a Both sides have a baryon number of 2. There are no leptons. The interaction is possible.   2 marks

b Left-hand side has mu lepton number −1, but right-hand side has mu lepton number 1, hence the interaction is not possible.   1 mark

c Left-hand side has baryon number 2, while right-hand side has baryon number 1. The interaction is not possible as baryon number is not conserved.   1 mark

Question 17 (6 marks)

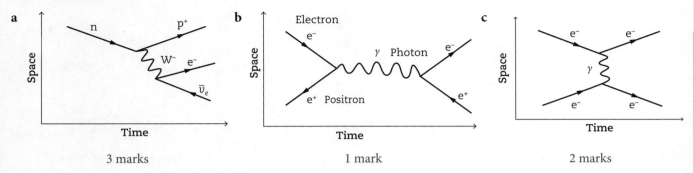

a 3 marks

b 1 mark

c 2 marks

Question 18 (4 marks)

Fermion generations

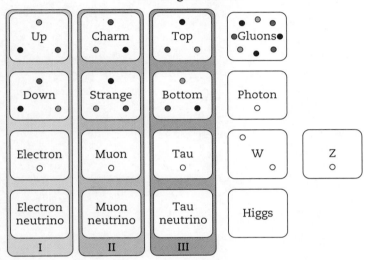

FIGURE 7.3 The Standard Model

Mark breakdown:

- 1 mark for correctly drawn quarks
- 1 mark for correctly drawn leptons
- 1 mark for correct gauge bosons
- 1 mark for correct format and setting out

Question 19 (4 marks)

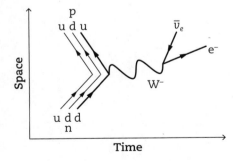

Mark breakdown:

- 1 mark for components and axes set out correctly
- 1 mark for quark nature of particles correctly identified (proton and neutron)
- 1 mark for correct directions or arrows on lines
- 1 mark for correct boson identified

Question 20 (4 marks)

a

Force	Gauge bosons
electromagnetic	photon
strong nuclear	gluon
weak nuclear	W and Z boson

3 marks

b The quarks, like the leptons, are fundamental particles as they are not composed of any other, smaller particles.

1 mark